Sustainable Development in the 21st Century

Volume 3

Jan-Niclas Gesenhues

Smart Energy in Mozambique

Drivers, Barriers and Options

Nomos

© Coverpicture: NASA Earth Observatory image by Joshua Stevens, using MODIS data from the Land Atmosphere Near real-time Capability for EOS (LANCE). Caption by Kathryn Hansen.

The Deutsche Nationalbibliothek lists this publication in the Deutsche Nationalbibliografie; detailed bibliographic data are available on the Internet at http://dnb.d-nb.de

a.t.: Münster, Univ., Diss., 2019

ISBN   978-3-8487-6562-1 (Print)
        978-3-7489-0679-7 (ePDF)

British Library Cataloguing-in-Publication Data
A catalogue record for this book is available from the British Library.

ISBN   978-3-8487-6562-1 (Print)
        978-3-7489-0679-7 (ePDF)

Library of Congress Cataloging-in-Publication Data
Gesenhues, Jan-Niclas
Smart Energy in Mozambique
Drivers, Barriers and Options
Jan-Niclas Gesenhues
283 pp.
Includes bibliographic references.

ISBN   978-3-8487-6562-1 (Print)
        978-3-7489-0679-7 (ePDF)

Onlineversion
Nomos eLibrary

1. Edition 2020
© Nomos Verlagsgesellschaft, Baden-Baden, Germany 2020. Printed and bound in Germany.

For Annica and Carlotta

Preface

This thesis was accepted as a dissertation at the University of Münster in the summer semester of 2019. It is particularly dedicated to the analysis of decentralized and intelligently networked energy sectors.

Countries around the world are undergoing a paradigm shift in energy supply – from centralized, fossil-fueled supply systems to a decentralized, intelligently networked and climate-friendly structure. Some countries in the global south play a key role in this development. Using Mozambique as an example, this study shows how a digitally networked energy supply system can grow "from below". On this basis, strategies are developed that can contribute to achieving some of the United Nations' Sustainable Development Goals - especially in the areas of energy, climate, health, economy and poverty reduction.

My special thanks go to my two supervisors Prof. Dr. Norbert Kersting and Prof. em. Dr. Paul Kevenhörster, for their scientific and moral support throughout the research process.

I am also very grateful to Prof. Dr. Boaventura Chongo Cuamba from Eduardo Mondlane University of Maputo for his support, expertise and networks. Our scientific cooperation led into a partnership project between Mozambican and German institutions, funded by the German Ministry of Economic Cooperation and Development. This project addresses the needs of the renewable energy sector in Mozambique and is a great opportunity to use the scienfic insights of this thesis in practice.

This work could not have been done without intensive investigations and expert discussions on site in Mozambique. I would, therefore, like to thank all respondents and express my gratitude to the Heinrich Böll Foundation and the German Academic Exchange Service (DAAD) for funding part of my field research in Mozambique and South Africa.

I was priviledged to develop my thesis together with an international group of PhD students with a strong expertise in development politics, digitalization and with much experience from East-African countries. I am especially grateful to my colleagues Phillip Hocks M.A., Dr. Andrew Matsiko and Lia Polotzek M.A. for reviewing the manuscript and for very helpful comments and discussions.

Münster, January 2020   *Jan-Niclas Gesenhues*

Contents

Contents

List of acronyms

AU	African Union
AC	Alternating current
ALER	Associação Lusófona de Energias Renováveis
AMER	Associação Moçambicana de Energias Renováveis
App	Application
ARENE	Autoridade Reguladora de Energia
CIA	Central Intelligence Agency
CNELEC	Conselho Nacional de Electricidade
DC	Direct current
EDM	Electricidade de Moçambique
EnDev	Energising Development Program
FDI	Foreign direct investment
FUNAE	Fundo da Energia, National Energy Fund of Mozambique
FRELIMO	Frente de Libertação de Moçambique
GDP	Gross domestic product
GIZ	Gesellschaft für Internationale Zusammenarbeit
GWh	Gigawatt hour
HCB	Hydroelectricity of Cahora Bassa
ICT	Information and communication technology
IMF	International Monetary Fund
INE	Instituto Nacional de Estatística
kV	Kilovolt
kWh	Kilowatt hour
MPD	Ministério de Planificação e Desinvolvimento
MZN	New Mozambican Metical
OAU	Organization of African Unity
OECD	Organization for Economic Co-operation and Development
PayGo	Pay-as-you-go technologies
RENAMO	Resistência Nacional Moçambicana
RSA	Republic of South Africa
SADC	South African Development Community
SASGI	South African Smart Grid Initiative
UN	United Nations

List of acronyms

UNCTAD   United Nations Conference on Trade and Development
WLAN   Wireless Local Area Network
ZANLA   Zimbabwe African National Liberation Army

List of symbols

C	Cost function
D	Demand function
ε	Price-elasticity of demand
mc	Marginal costs
mr	Marginal revenue
n	Sample size
p	Price
p_o	Off-peak-price
p_p	Peak-price
π	Profit
R	Revenue
sd	Standard deviation
u	Utility
μ	Average value
x^D	Demanded quantity of the commodity
x_i	Quantity of the commodity i
x_o	Off-peak quantity
x_p	Peak quantity
x^S	Supplied quantity of the commodity
y	Number of clients

1. Introduction

According to data provided by the International Energy Agency, about two thirds of the Mozambican population do not even have basic access to electricity (IEA 2017). That is, broadly 20 million Mozambicans are not or not regularly supplied by electric energy (UN 2018). Even households, connected to the grid, can often not rely on a stable energy supply. The Mozambican power transport and distribution system faces high transmission losses and lacks load management. Clients are hit by regular blackouts and the electricity sector is characterized by inefficiencies. The resulting lack of access to reliable electricity is a severe barrier to human and economic development in Mozambique.

Smart energy is proposed as a technology, expected to significantly improve the reliability, efficiency and environmental friendliness of power supply in Africa (Welsch et al. 2013). Energy infrastructure is considered "smart" in this study if it is thoroughly interconnected by information and communication technology (von Lucke 2017, 158) in order to improve the generation, transmission, distribution and management of energy flows (EPRI 2011, 1). Intelligently interconnected energy systems are expected to be able to contribute substantially to human development, quality of living and to the economic attractivity of a region (von Lucke 2017, 158). Thus, smart energy could help achieving the United Nations´ Sustainable Development Goals as goals seven and 13 explicitly demand access to affordable modern energy, a reliable and sustainable energy supply and the mitigation of climate change (UN 2015).

In Mozambique, first smart grid projects have already been launched. For example, large industrial consumers were equipped with smart meters and some generation capacities are already monitored and optimized using information and communication technology. Smart mini grids and innovative energy systems without any grid access, so-called off-grid solutions, have been increasingly implemented in the last years to relieve the severe lack of access to electricity in Mozambique´s rural areas.

The emergence of smart energy solutions is one effect of digitalization as an encompassing technological, social and economic transformation. In general, digitalization is proceeding also and particularly in countries of the global south like Mozambique. Some digital innovations are applied especially in the global south, such as intelligent off-grid energy systems, mobile payment solutions or certain participatory innovations to monitor and influence political processes (Tenenbaum et al. 2014, 20 ff., Shayo, Kersting 2016, Shayo, Kersting 2017, Kersting, Matsiko 2018). When industrialized countries pursue a digitalization strategy of upgrading existing communication, transport, production or energy supply infrastructure, countries in the global south can directly implement the most recent digital technologies as no or only little equivalent infrastructure has existed so far. That is, in regions without access to electricity – which until today are frequent in Mozambique – conventional grid solutions could possibly be leapfrogged and a smart power system installed from the outset. However, despite first approaches, Mozambique is still far away from a deeply digitalized energy sector, such that there is still a long way to go.

Nevertheless, the existing conditions and developments in Mozambique form an interesting framework to analyze smart and innovative electrification strategies. In Mozambique, the implementation of innovative solutions for the generation and distribution of electricity can be observed under real-world conditions. While most of the industrial countries worldwide are currently trying to diversify and decentralize their mainly centralized and fossil-dependent energy supply, Mozambique is already applying central and decentral smart solutions simultaneously – often based on renewable energy sources. Therefore, some of the experience from developing countries like Mozambique can be quite valuable for industrialized countries with a mature but highly centralized and pollution-intensive energy supply. Consequently, analyzing the implementation of innovative and smart energy solutions in Mozambique can deliver valuable insights to develop successful strategies for a cleaner, more reliable, affordable and smarter energy supply – not only for Mozambique and its neighboring countries in East Africa but also beyond.

The analysis of smart energy innovations and their diffusion is a manifold task. Central actors and the given framework conditions of innovation diffusion must be examined while questions like the following arise: Which resources are available for the innovation process? How exactly can information and communication technology be involved? Which political, economic, technological and social influences shape the innovation process? Which internal and external determinants exists? Finally, how sustainably and durably can the innovation be expected to be applied (Kersting 2017, 2). That is, if Mozambique´s future energy sector will be smarter and more effective, depends on the specific economic, technological, political and social drivers and barriers and their impact on the different options for a smart electrification.

In Mozambique, comprehensive research about these drivers and barriers to smart energy and to electrification in general has not yet been conducted. Experience from other countries cannot be transferred to the specific context of the electricity sector in Mozambique without severe threats to validity. For the same reason, transferring experience from conventional grid systems to smart grids is not advisable, either. Therefore, specific research about the drivers and barriers to electrification and smart energy in Mozambique is promising to deliver new insights about opportunities and challenges in electrification and smart energy implementation. This study aims to reveal, prioritize and explain these drivers and barriers. The corresponding part of this doctoral thesis, dealing with the drivers and barriers to smart grids in Mozambique, refines, extends and deepens a preliminary analysis of the same topic, developed in the author´s master thesis.[1]

Based on the revealed drivers and barriers, this study aims to analyze in a second step which smart energy solutions are most suitable for the Mozambican context. That is, the point of interest is which smart energy solutions most effectively cope with the given barriers and benefit from the given drivers. Different options for a smart electrification are for example: a large interconnected and intelligent grid that covers nearly all parts of the country, isolated or interconnected smart mini grids or smart off-grid solutions. Due to an unsatisfying pace of grid extension, limited financial resources and a dispersed population in most parts of the country, intelligent regional grids and smart off-grid solutions have recently attracted more and more attention in electrification strategies for Africa and Mozambique in particular (Blyden, Lee 2006; Tenenbaum et al. 2014; Cescon 2015; GreenLight Consult 2016, DFID 2016; Sampablo et al. 2017).

Electrification in Mozambique is a pressing issue but not all barriers can be overcome in the short term. Most of them are structural shortcomings in the political and economic system of the country and require a strong motivation, effective coping strategies and a large amount of endurance. To achieve a significant progress in electrification in a short term, it is therefore reasonable to analyze which electrification solutions

---

1   See Gesenhues (2016).

can be realized best under the given conditions instead of trying to improve power supply by tackling the whole battery of structural barriers. Hence, the question of this study is which smart electrification solution is most promising under the given drivers and barriers.

To address the questions, raised above, it is regarded as necessary to follow an interdisciplinary approach. The topic of drivers and barriers to electrification and the analysis of smart energy options have political, economic, technological and cultural dimensions. Therefore, a limitation of the analysis to approaches from a single scientific discipline should be avoided. As Tenenbaum et al. (2014, 313) put it: "When one has expertise or experience in a particular area – whether is its engineering, economics, marketing, law, regulation, or another field – there is a natural tendency to define key problems and solutions in terms of one´s expertise. It is important to resist this temptation."

The remainder of this thesis is organized as follows: Chapter 2 gives a brief overview of the Mozambican electricity sector. In chapter 3, some vital potentials of a smart power system for Mozambique are presented. The theoretical foundations of innovation diffusion are presented in chapter 4. Whether a new innovation will be implemented in the Mozambican power sector depends on how well and how easily it diffuses through the market and its surrounding social system. Therefore, it is important to understand which factors determine the pace and extent of the adoption and implementation of innovations. Potential drivers and barriers for smart grids and electrification will be derived from the determinants of innovation diffusion, postulated by the diffusion theory.

Chapter 5 takes a closer look at the drivers and barriers to smart energy in Mozambique. Chapter 5.1 introduces the methodology of this study, describes the research process, addresses methodological challenges and derives potential drivers and barriers to a smart electrification in Mozambique. As a result, the empirical tools – a questionnaire and an interview guide for the qualitative interviews – are developed. Chapter 5.2 starts with the main findings of the empirical analysis. In the main findings, a statistical analysis and a qualitative evaluation of the empirical results reveal and prioritize the drivers and barriers. Subsequently, in chapters 5.3 to 5.6, the particular drivers and barriers are described and explained in detail. A special focus rests on the analysis of the theoretical foundations, the origins, interconnections and impacts of the variables that form the drivers and barriers. The analysis enables implications of how decision makers from politics or business could address the identified drivers and barriers. Chapter 5.3 starts with the ability and the willingness to pay for smart grids and power infrastructure in general, 5.4 analyzes the economic and political conditions in the Mozambican electricity market and 5.5 focuses on the applicability, the management and the economic feasibility of a potential smart grid infrastructure in Mozambique. Chapter 5.6 examines the characteristics of governance and the political environment for electrification in Mozambique. The quality of institutions, decision-making and regulation are analyzed. Furthermore, the effects of international cooperation and development assistance on the Mozambican power sector are examined. Political cleavages and conflicts that shape the economic development in Mozambique are described, including their political and historical origins.

The analysis of drivers and barriers in all sub-chapters of chapter 5 includes approaches for further research, in case that future studies aim to analyze certain drivers or barriers in even more depth than it can be presented in this study. Chapter 6 presents preliminary conclusions for the analysis of drivers and barriers and connects it with the following analysis of the drivers´ and barriers´ impact on different options for a smart electrification.

The second part of this thesis analyses how the different options for a smart and effective electrification perform under the given drivers and barriers. The goal is to reveal which solution or which solutions are expected to be most suitable for the Mozambican needs and characteristics. Chapter 7 portrays the central and the decentral approach to electrification and introduces the corresponding technological solutions, such as main grid extension for the central approach and isolated mini grids, connected mini grids and off-grid solutions for the decentral approach. In chapter 8, for each option, it is examined how it performs under the drivers and barriers conducted in the first part of this study. Eventually, each option´s feasibility, given the framework conditions in the Mozambican energy sector, is assessed. Chapter 8 finishes with concluding remarks for the different options´ feasibility and aims to determine to which extend, under which conditions and in which regions of the country each option can be expected to be beneficially implemented. The several recommendations for political reforms, proposed by the respondents during field researches, aiming at improving the conditions for smart electrification in Mozambique, are summarized and evaluated in chapter 9. The last chapter summarizes the most important findings of this study and presents an outlook towards Mozambique´s future – potentially smarter – energy supply.

In total, 31 experts and actors from the Mozambican energy sector provided first-hand information for this study. The answers of 23 respondents to relatively short questionnaires formed the ground for the deeper qualitative research and supplemented the following expert interviews. Insights from 18 comprehensive qualitative interviews with representatives from the different areas of Mozambique´s energy sector, are the central empirical pillar of the following analysis. The empirical insights will be supplemented by comprehensive theoretical reasoning based on the existing literature about the topics of interest.

2. The Mozambican electricity sector

2.1. Basics of the Mozambican electricity sector

The Mozambican government set out the goal to pursue electrification, as to 50% of the households should be connected to the grid in 2023 (World Bank 2015, 7). To achieve this goal, significant acceleration of electrification is necessary as broadly 70% of the population remain without access to a basic electricity supply until today. Including electricity supply from alternatives to the main grid, such as mini-grids and standalone systems, electricity access in Mozambique increased from 7% in 2000 and 2005 to 15% in 2009 and 29% in 2016 (IEA 2017). In these International Energy Agency (IEA) numbers for electricity access, households are included which have access to electricity supply that satisfies basic modern energy needs. This lower bound of electricity access is defined by the IEA as sufficient electricity – no matter what source – to operate four lightbulbs at five hours a day, one refrigerator, a fan operating six hours daily, a mobile phone charger and a television, operating four hours daily. This bundle of electricity equals a yearly electricity consumption of 1250 kWh per household with standard appliances and 420 kWh with efficient appliances (IEA 2018).

It is reasonable to assume that access some source of electricity in Mozambique is actually higher than the 29%, measured by the IEA, as for instance, a solar panel with two lightbulbs and a phone charger is also some access to energy but it falls below the bottom line of modern energy access as defined by the IEA (see above). Households, not included in the IEA numbers, only have little electricity access which is insufficient to simultaneously satisfy the basic energy needs mentioned above such as radio, television, lighting and cooling.

Although some progress in electrification has been made in the last years, the vast majority of Mozambicans still lives without access to enough electricity to satisfy basic energy needs. Besides, even the Mozambicans who have grid access are hit by regular blackouts and power distribution suffers from significant transmission losses and energy theft, originating from illegal connections.

Access to electricity in Mozambique reflects a sharp urban-rural and north-south disparity. While in 2016, 57% of urban households had electricity access, it was only 15% of the rural ones (IEA 2017). The north-south disparity becomes apparent, considering that about 53% of the power consumption takes place in the southern provinces, 21% in the central and only 13% in the north (remaining 13%: exports and special customers, e.g. aluminum smelters) (EDM 2013, 42). The general north-south and urban-rural disparity remains relatively stable as the extension of energy supply in Mozambique proceeds only at a moderate pace and does not reflect a strong focus on the so-far neglected areas (ALER 2017, 88 f.).

After this introduction to the output of the Mozambican electricity sector, we shall take a closer look at the actors. Electricidade de Moçambique (EDM) is the national power utility, founded in 1977 as a public company, uniting the different until then existing energy suppliers. Engaged in generation, transport, distribution and commercializing of electricity, EDM is a fully vertically integrated state-owned utility (Government of Mozambique 2009) which is undoubtedly the most powerful actor in the Mozambican electricity market.

The second important actor in the Mozambican energy sector is the National Energy Fund (FUNAE). The initial intention was to create a fund that concentrates on financing and guaranteeing innovative public and private energy projects, contributing especially to the development, production and implementation of alternative forms of energy supply (Government of Mozambique 2009). However, due to EDM´s reluctance to renewable energies and off-grid electrification, FUNAE has become continuously more engaged in the operating business of these areas.

Mozambique´s energy market offers a still small but growing private sector, consisting mainly of small and medium enterprises engaged in renewable energies, mini grids and off-grid electrification. Such decentral supply can bring energy to so far unsupplied areas. Recent technological advancements have made off-grid and mini grid systems more feasible. Most of the companies from the decentral and renewable market are represented in the Mozambican Association of Renewable Energies *AMER* which promotes the interests of this sector and was founded in 2017 (AMER 2019).

In the government, the Ministry of Mineral Resources and Energy executes and proposes legislation for the energy sector. Its headquarters are located in the country´s capital Maputo and it holds regional delegations in each of the provinces. Power tariffs are set by the ministry. Until recently, the National Council for Electricity (CNELEC) which consisted of experts and high officials, advised the ministry and the council of ministers on tariffs and concessions (World Bank 2015, 18). A legislative process has been completed to further develop CNELEC to be Mozambique´s fully independent energy regulator (Sampablo et al. 2017, 3). Eventually, in late 2017, the Mozambican parliament passed a bill creating CNELEC´s follow-up organization, the Energy Regulatory Agency (ARENE) which has started operating in 2018. ARENE faces high expectations from stakeholders to be a capable and politically independent regulator, setting tariffs and the regulatory framework for Mozambique´s energy sector (ALER 2018).

2.2. *Generation of electricity*

Total electricity generation in Mozambique increased substantially between 2003 and 2016 from broadly 11 TWh to approximately 18.75 TWh (ALER 2017, 76). As already stated before, a relevant part – about 13% of total generation (EDM 2013, 42) – is exported. By far the largest generation in Mozambique comes from hydro power and is especially taking place at the Cahora Bassa Dam in the province of Tete at Zambeze river. The Dam is managed by Hydroelectrics of Cahora Bassa (HCB) and provides five turbines of 415 MW each, resulting in a total capacity of 2,075 MW (Government of Mozambique 2009, 9). With 92.5% of the shares, EDM is the largest shareholder of HCB (World Bank 2015, 19). About 85% of the electricity, sold by EDM originates from the Cahora Bassa dam, 6% are supplied by other EDM-owned generation facilities and 2% from independent domestic producers. The remaining gap is filled by imports (EDM 2013, 9).

Due to a vast abundance of primary energy sources, Mozambique possesses large potentials to increase its power production. As a subtropical country, close to the equator, Mozambique is one of the world´s most viable countries for solar power (Blyden, Lee 2006, 2, Sampablo et al. 2017, 28). Especially the coastal areas also have large potentials for wind use (Blyden, Lee 2006, 3). According to Sampablo et al. (2017, 30), wind resources sum up to a potential of 230 MW of high potential sites.

Looking at hydro power, the currently planned Mphanda Nkuwa Dam which shall be located on Zambeze river, further downstream of the Cahora Bassa dam, is the current flagship project in Mozambique. It has a planned capacity of 1500 MW (World Bank 2018 a). However, the Mphanda Nkuwa project is highly controversial due to vast resettlement, environmental impact and the worries that the generated energy will mainly be sold to South Africa (Machena, Maposa 2013, 6).

On the fossil side, Mozambique´s offshore gas reserves are estimated at 277 trillion cubic feet and estimated offshore coal reserves sum up to more than 20 trillion tons (Cipriano et al. 2015). Exploration and exploitation of offshore-resources are recently increasing, such that deposits might be even much larger than currently assumed. Additionally, proven onshore natural gas reserves of approximately 3.5 trillion cubic feet add up to the resource base (ibid.). Gas-fired power generation is on the rise in Mozambique, slowly narrowing Cahora Bassa´s market share (World Bank 2015, 24).

2.3. Transmission and distribution

As the previous chapters have already shown, the clear majority of the Mozambicans is not connected to the power grid. Customers are additionally hit by regular blackouts and load shedding. Transmission losses and power theft further impair the quality of the Mozambican electricity supply (ALER, 2017, 78). So far, transmission lines only reach the main cities. Towns situated along the lines are supplied as well. Rural electrification in general remains at a low level, though. Centralized generation, large geographical distances and a low density of population make long transmission lines necessary which increases transmission losses and brings about further threats of instability, especially at the outskirts of the grid. Especially in remote areas which are not connected to the grid, diesel generators (Ahlborg, Hammar 2014, 118), charcoal or firewood are usually the only available sources of energy. Still today, more than 80% of Mozambique´s primary energy is delivered by biomass (Veremachi et al. 2016, 1).

Mozambique´s grid network consists of three separate systems (ALER 2017, 85): a northern, a central and a southern, while there are some minor connections between the central and northern system. Figure 1 illustrates Mozambique´s grid systems to date.

As the map shows, the southern grid system sustains Maputo and reaches out to the neighboring countries. Most of the electricity, generated at the Cahora Bassa hydropower plant is exported to South Africa. The

Maputo-region (southern system) receives a large share of its power from the Cahora Bassa plant as re-imports via South Africa since no direct connection exists.

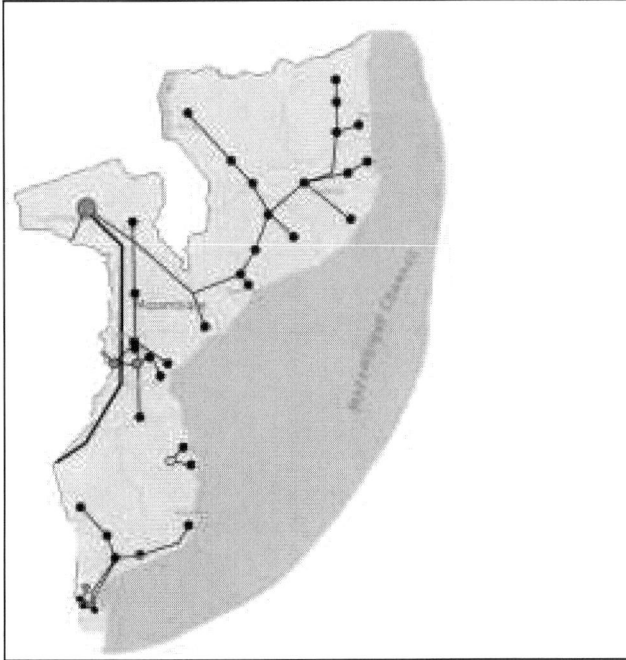

Figure 1: *The Mozambican electricity grid.*
Source: Own illustration based on World Bank (2015, 19).

While the northern system is also mainly fed by Cahora Bassa, the central system relies on local generation, also mainly hydropower. Since there is only one line from Cahora Bassa to the north, the northern system is highly vulnerable if problems with the connection to Cahora Bassa occur. Except for the two 535 kV extra-high voltage lines transporting power from Cahora Bassa to South Africa, Mozambican power lines reach a voltage of no more than 275 kV (EDM 2013, ALER 2017, 84 f.)[2]. The connection from Cahora Bassa to the northern system is regularly overloaded which –

---

2   For more information about the voltages, the interconnections, substations, dispatch centers and newly planned power lines see EDM (2013).

alongside with further capacity and maintenance problems in the grid – leads to load shedding, transmission losses and disconnections of households for several hours daily (World Bank 2015, ALER 2017, 78 f.).

Facing the problems of transmission and distribution of electrical power, the Mozambican Government acknowledges that the grid is a distinct bottleneck to the electrification of the country and requests EDM to extend the grid to all districts (Government of Mozambique 2009). Thus, the government's goal to achieve 50%-access by 2023 (see above) shall be achieved. EDM connects approximately 118,000 households each year (World Bank 2015, 7). Furthermore, new transmission lines are planned and constructed to improve the interconnections between north, central and south[3]. The flag-ship-project is a new system of extra high and high voltage lines from the center to the south, where a strong direct connection between Cahora Bassa and the consumption-intensive South should be accomplished (EDM 2013). Efforts of EDM are supplemented by FUNAE and independent actors such as small and medium companies which further pursue off-grid and mini-grid-electrification (World Bank 2015, 20, ALER 2017, 78).

To address the lack of reliability and resilience of the grid, smart technologies could be of support. For example, EDM has implemented a roll-out-program of advanced metering infrastructure for its 4000 largest users. The utility aims to improve forecasting and load management by systematic recording and monitoring of consumption profiles (World Bank 2015, 10). Also, for decentral energy supply infrastructures, smart solutions are already operating in Mozambique and can alleviate some of the sector's problems, as following chapters will show.

---

3   A detailed map of the planned power lines can be found in the preface of EDM (2013).

3. Potentials of smart energy for Mozambique´s electricity sector

For decades, electricity grids were designed for a one-way process: the electricity was generated centrally in large plants and distributed through a large grid from the generator to the consumer. This passive character of the demand side has changed radically. With increasing levels of decentral generation and storage capacities, accompanied by new technological solutions for demand response, electricity transmission and distribution have become a two-way process of constant exchange of electricity and information between the consumer side and the supply side.

In advanced electricity markets, there is no pure consumer side any more. With increasing levels of decentral generation, like rooftop photovoltaics or on-site diesel generators, households and businesses have become consumers and producers ("prosumers") at the same time. In most cases, on-site generation does not equal on-site consumption such that lacking energy has to be taken from the grid or surplus energy fed into the grid. To manage the increased participation of the demand side and an increasingly distributed generation with a large proportion of decentral, fluctuating renewable energies, there is enhanced need for co-ordination. At the same time, this radical change in electricity generation and distribution comes with new potentials to increase reliability, efficiency and sustainability of the electricity sector. To manage the challenges and to exploit the potentials of these changes in modern electricity sectors, several new technologies and concepts were introduced, summarized under the term "smart grid" (Sioshansi 2012).

A comprehensive understanding of the concept of a "smart grid" requires the definition of the term "smart". In here, an infrastructure is referred to as "smart" if it is thoroughly interconnected by information and communication technology. Intelligently interconnected infrastructures such as buildings, factories, authorities, cities or power grids can contribute substantially to human development, quality of living and to the attractivity of a location (von Lucke 2017, 158 f.).

Consequently, a "smart grid" describes to a thoroughly modernized electricity system which monitors demand and supply patterns and optimizes the electricity supply-system by automatically coordinating its interconnected elements – generation, transmission, distribution, consumers, storage installations and end-use-devices (EPRI 2011, 1). Sensors and smart meters collect information which is subsequently transmitted by information- and communication technology (e.g. data lines, mobile communication systems, power-line communication, radio-services) to evaluation technologies. Data evaluation allows for effective response measures (e.g. load management) by the grid operator, using inter alia data aggregation and evaluation software, energy-management systems, load-management systems, meter-data management-systems and distribution-management systems (Dada 2014, 1007 f.). That is, a smart grid follows a four-step architecture – data-collection, data-transmission, data-evaluation and response.

A smart energy supply system is an important building block of a smart city or more generally spoken, a smart region. A smart city or a smart region aims at improving the quality and efficiency of services for the local community with ICT-based solutions (Kersting, Zhu 2018, 256). Besides the local energy supply infrastructure, such services are for instance transportation systems, waste management, schools, libraries, administrative services or possibilities for an enhanced citizen participation in local politics (ibid.). The result is urban or regional innovation which consists of projects and programs that aim to change the local structures and procedures. In this definition, innovation is not per se to be regarded as positive or negative but rather neutral in its outcome (Kersting 2017, 1).

The automatic harmonization of energy demand and supply and the optimization of the electricity system by a smart grid facilitates the integration of fluctuating renewable energies, reduces transmission losses and improves demand management, so that outages and load-shedding are avoided. By a two-way flow of information and a real-time automatic demand response, a smart energy system does not only adjust supply to demand (like in a conventional supply chain) but manages demand and supply in a mutually adjusting process (Crispim et al. 2014, 92). That is, smart energy systems recognize non time-critical energy consumption and allocate the energy according to the current availability. For instance, for an electric car or a phone which should be charged during the night, it does not matter when exactly it is charged as long as it is fully charged the next morning. A smart energy system recognizes these flexibility options and uses them to reduce peak demand – in this example, the car or phone battery are charged when energy is easily available but cut off if shortages occur. That is, what mainly distinguishes a smart grid from a conventional one, is its potential of self-healing, participation of customers and automatic two-way-management of demand and supply (Dada 2014, 1008).

Some but certainly not all the concrete advancements enabled by smart energy solutions are the following (Sioshansi 2012):

- Better integration of distributed and decentral generation sources, including fluctuating renewable energies.
- Enabling intelligent devices to adjust usage based on differences in the electricity price and scarcity.
- Better integration and participation of central and decentral storage devices to intelligently store energy when it is available and utilize it when the system runs short of power.
- Allow decentral generation, demand-response options and decentral storage to actively participate in balancing the load.
- Make the grid more robust and more reliable by measures of self-healing.
- Induce savings by better operation and maintenance of the network.

Especially the combination of an improved demand side management, a more sophisticated control and forecasting as well as a more effective prioritization of loads is promising for the Mozambican context since these attributes can help to reduce the severe distribution losses, outages and load shedding. A better control of power flows can additionally help to detect and reduce power theft which is an important problem in Mozambique (see chapter 0).

The vast potentials for decentral renewable energies in Mozambique can be exploited more effectively if their fluctuations are reliably buffered by an intelligent power grid. A positive side effect of a stronger use of renewables is the possibility to substitute emissions-intensive energy sources and expensive generator solutions – commonly used in rural Mozambican areas without access to other energy sources. The reduced need of fossil reserves means more environmental sustainability of power generation, more intergenerational justice and a contribution to the reduction of greenhouse gases. The mitigation of global warming is of high importance for Mozambique since the country is extremely vulnerable to the effects of climate change. A large proportion of the Mozambicans lives in low-lying coastal areas. The infrastructure is mainly not climate-resilient with houses made from basic natural materials (wood, loam, plants) still being the dominant picture in most rural areas. Furthermore, agriculture and fisheries – the main source of income and food for a large part of the population – are extremely vulnerable to climate change. The destructive floods and droughts which already today are regular events in Mozambique, can be expected to become even more frequent if climate change proceeds (USAID 2013).

Besides, a smart energy supply can become an important locational factor for the Mozambican economy. The reduction of losses from power theft or transmission problems, a more efficient power sector and the possibilities for a more diversified power generation help to keep energy costs low. A reliable energy supply that comes at low costs is one of the most important drivers for economic development. A better power supply furthermore improves the well-being of the people, and enhances the quality of social services such as education and health.

Since the extension of the central grid is unlikely to deliver electricity to the majority of Mozambique's population in the short and medium future, decentral options for electrification such as isolated mini grids or off-grid solutions are increasingly considered to fill the gaps. With the new possibilities for load management even in small distribution systems, these decentral approaches to electrification become more effective and more viable through smart solutions. That is, not only large grid networks can be smart, but also isolated mini grids or even off-grid solutions can apply information and communication technology for improved monitoring, load management, billing or remote control (see chapter 7). Thus, smart energy opens new chances for decentral and rural electrification. In the following chapters, grid-based and off-grid smart energy technologies will be summarized as "smart energy solutions".

Apart from the specific potentials for the power sector, smart energy in Mozambique comes with an additional advantage if one looks at the possibilities for their implementation. Since several new power lines are planned in Mozambique (see ch. 2.3), the new grid-systems could directly be smart ones. Thus, Mozambique could leapfrog conventional supply technologies in some regions. Leapfrogging describes the process when the adopter of an innovation leaves out some stages of technological development and directly implements a new, more advanced technology (Fudenberg et al. 1983). Advancements of existing technologies do not necessarily only improve the existing technology but also offer possibilities which the older technology did not offer at all. These new qualities might lead additional individuals or organizations to adopt the new technology although they did not adopt the preceding technology. Thus, they leapfrog the older technology and directly start with the new one (Fudenberg et al. 1983, 3). Leapfrogging countries benefit from technological developments from other parts of the world and thus, catch up in economic development (Murphy 2001).

In a brief summary, smart energy in Mozambique has the potential to lead to improvements in all of the three core goals of energy politics: reliability, low costs and environmental sustainability. The energy security can be enhanced, the energy sector can be more efficient, and the potentials of fluctuating renewables can be exploited more extensively. A shift from central to decentral generation, distribution and consumption is facilitated by smart solutions. Interruptions, transmission losses, peak loads and power theft can be reduced by more flexibility and by improved control of energy flows. As a result, a smart energy sector can improve the services for the customers and be an important locational factor for business. Furthermore, leapfrogging potentials can be an additional argument for the implementation of smart energy in Mozambique.

4. Diffusion of an innovation

It is the focus of this study to analyze the diffusion of an innovation – smart energy – in the Mozambican power market. To determine which aspects are important to look at in this analysis, one has to understand what shapes the diffusion process of an innovation, what hinders and what supports the sustainable adoption of a new technology. This chapter presents a theoretical framework for typical innovation diffusion processes with a special emphasis on the diffusion of technologies in organizations, such as energy companies.

The probably most influential theory on the diffusion of innovations was developed by *Everett M. Rogers* and portrayed in his book "Diffusion of Innovations", published in the first edition in 1962. Since then, most research on innovation diffusion has built on Rogers´ theory. In his book, Rogers extensively describes the underlying dynamics of diffusion processes. Most of the further scientific contributions focus on extending and refining Rogers´ theory, especially in the field of formal analytical modelling of diffusion processes.[4]

It is the purpose of this chapter to present a descriptive analysis of the innovation diffusion process to generate a basic understanding. Formal models of diffusion are not considered extensively. Instead, Rogers´ basic theory, supplemented by some contributions from further research which allow for helpful extensions of the basic theory.

In the remainder of this chapter, it will become evident that there are certain drivers and barriers which determine and shape the diffusion and adoption of innovations. Based on this theoretical analysis of general determinants of diffusion and adoption processes, potential smart grid-specific drivers and barriers will be derived in the proceedings of this study. These potential smart energy-specific drivers and barriers shape the ground of the analytical framework for the subsequent empirical field research.

---

4   For the formalization of Rogers´ theory and further reading see King 1963, Frank et al. 1964, Silk 1966, Arndt 1967, Robertson 1967, Bass 1969.

4.1. Principles of innovation diffusion

Rogers defines an innovation as an "idea, practice or object that is perceived as new by an individual or another unit of adoption" (Rogers 2003, preface). The innovation is usually a new means aiming at solving a certain problem (ibid.). A successfully established innovation creates a new industry or transforms an existing one (Senge 2006, 5). This transformation process is described by the term "creative destruction" (Schumpeter 1912): The structure of an economy is continuously revolutionized by innovations which destroy existing markets and create new ones. Creative destruction is a necessary process for economic and technological progress. The introduction of innovations like a new and revolutionary technology can enable new companies to dominate a new or transformed market. Thus, creative destruction has the potential to restructure an existing market entirely (Schumpeter 1912, Schumpeter 2005). If one thinks of the revolutionary economic and societal transformation, induced by the invention of the internet, the power of creative destruction becomes evident. For example, if smart energy solutions replaced conventional grid infrastructure in Mozambique, the conventional Mozambican energy market would be deeply transformed by a process of creative destruction.

For an innovation to induce a process of creative destruction, the bare invention of a new technology is not sufficient. The new technology has to diffuse, that is, it has to be "communicated trough certain channels over time among members of a social system" (Rogers 2003, 5). The result of communicating an innovation can be its adoption which is "the *decision* to make full use of an innovation" (Rogers 2003, 473). An innovation is considered as *implemented* if an adopting unit puts the innovation into use (Rogers 2003, 474).

The high importance of communication in the diffusion process becomes evident if one considers that every new idea, technology or practice comes with a large amount of uncertainty because reliable experience with the new solution does not exist at the time of its invention. In spite of that, a potential adopter has to be convinced that the innovation can effectively solve a certain problem, he or she perceives to have. An individual will only be convinced of adopting an innovation, once uncertainty is reduced to a tolerable level. Reducing uncertainty requires collecting information. That is, the innovation decision process is basically a process of information-seeking to reduce uncertainty about the characteristics of the innovation (Rogers 2003, 14). This study contributes to reducing the uncertainty about smart energy in the Mozambican power sector and enables a more informed decision about their potential adoption.

The pace of diffusion can be measured by the rate of adoption. The rate of adoption is "the relative speed with which an innovation is adopted by members of a social system. It is generally measured as the number of individuals who adopt a new idea in a specific period, such as a year" (Rogers 2003, 221). A graphic display of the rate of adoption is a helpful instrument to analyze the characteristics of a certain diffusion process. Typical diffusion processes create an S-shaped curve when the cumulative number of adopters is plotted against time (Rogers 2003, 11). Figure 2 gives an example for the process of a typical innovation diffusion.

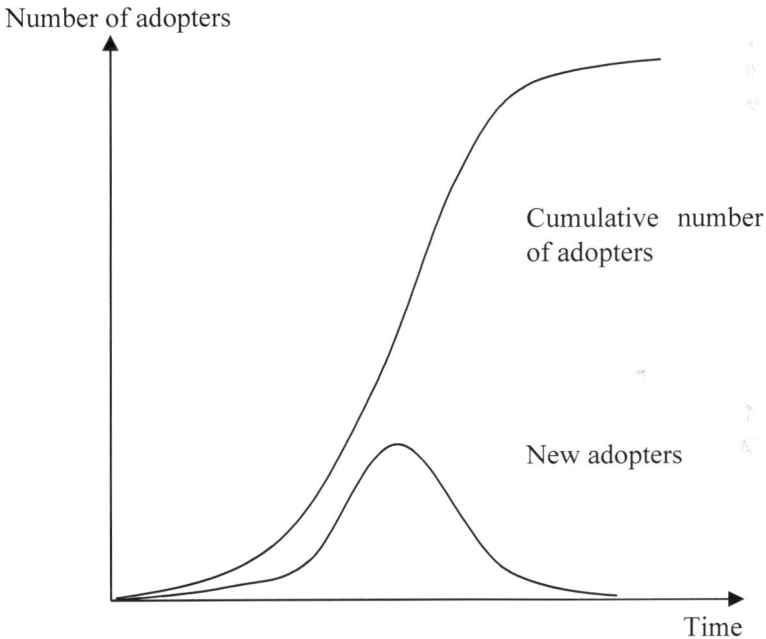

Figure 2: *Typical process of innovation adoption over time.*

Source: Own illustration based on Ryan and Gross (1943) and Rogers (2003, 273).

Figure 2 shows that the typical S-shaped curve of the cumulative number of new adopters corresponds with a frequency distribution of the number of new adopters per time period which approaches a normal, bell-shaped curve (Ryan, Gross 1943, Rogers 2003, 273). As in figure 2, a typical diffusion process starts with a low rate of adoption which represents a small number of pioneer adopters. Once a critical mass of adopters has been achieved (usually 10-20% of total adoption (Rogers 2003, 12)), the take-off phase with quickly increasing adoption follows. Here, the curve is steepest. Later, the curve becomes less steep again, entering the phases of late adoption and saturation of the market (Rogers 2003, 11). Diffusion processes of different innovations deviate by the slope and the precise shape of the curve. Some curves are steeper, representing a faster diffusion, some are less steep, representing a slower diffusion. Some innovations take longer until they reach take-off and some are quicker. Furthermore, innovations differ by the percentage of the members of a social system who have eventually adopted the innovation.

Based on the timing of adoption, the following classes of adopters can be specified: Innovators, early adopters, early majority, late majority and laggards (Bass 1969). It is reasonable to assume that later adopters are influenced by earlier adopters. The more individuals become adopters of an innovation, the more pressure arises for the remaining individuals to adopt the innovation, too. The only individuals who are not influenced by earlier adopters are the innovators. Unlike innovators, the remaining classes of adopters are all *imitators* (Bass 1969, 216). The innovators place the innovation into the social system and lead other individuals to imitate their adoption. Hence, the whole process of diffusion starts with and depends on the innovators (Bass 1969, 217).

The diffusion process is shaped by the social system, it takes place in. The rate of adoption is determined by factors such as the quality of communication channels (for example, mouth-to-mouth is slower than effective online marketing), the norms, habits, roles of opinion leaders and traditions of a society (Rogers 2003, 23 f.). These determinants have an impact on the acceptance and the willingness to pay for innovations and therefore influence their adoption (Mahajan et al. 1991, 164). For example, a very technology-skeptical society can be expected to show a lower rate of diffusion for smart grids than a relatively technology-optimistic society. Further conditions which shape the adoption and implementation of innovations are spatial and environmental framework conditions (Mahajan et al. 1991, 164). Usually, cities or urban structures are described as favorable environments for the development of innovative procedures and technologies. Creativity, technological openness and heterogeneity contribute to an enabling innovation environment in cities and urban structures (Kersting 2017, 1).

Apart from the characteristics of the social system, the rate of adoption is especially determined by the innovation itself. To put it more precisely, what matters are the *perceived* properties of an innovation: The decision to adopt an innovation depends on the subjective perceptions of the innovation by individuals not on objectively verified characteristics (Rogers 2003, 223). Even if a new product has proven to be completely ineffective in scientific trial, individuals might still adopt it if marketing makes them believe that the product has certain beneficial attributes for them.

The implication that the innovation decision is probably mainly driven by perceptions, not by objective attributes, makes technology diffusion a social process. This insight makes diffusion research a matter of social science. To evaluate whether an innovation can be expected to diffuse successfully in a certain environment, one has to reveal the relevant actors´ perceptions about this innovation, its drivers and barriers.

The perceived quality of an innovation is a combination of attributes which are observable in the present and by expectations about unobservable attributes and future developments of the product (Mahajan 1991, 160). For example, if a consumer expects the price of a product to fall in the future, the consumer might be reluctant to purchasing the product in the present.

According to Rogers (2003), the core determinants for the rate of adoption among an innovation´s perceived attributes are: Relative advantage, compatibility, complexity, trialability and observability. What quality smart energy solutions show regarding these attributes, determines their chance of a successful diffusion and implementation. In the following, the mentioned determinants shall be described in detail.

- **Relative advantage:** The relative advantage of an innovation is defined as "the degree to which an innovation is perceived as better than the idea it supersedes" (Rogers 2003, 15). Hence, the relative advantage is "a ratio of the expected benefit and the costs of adoption of an innovation" (Rogers 2003, 223). The cost-benefit ratio increases in favor of the new innovation with its profitability, low initial costs, a decrease in discomfort, social prestige, savings in time and effort and immediacy of reward. Missing immediacy of reward is one reason why preventative innovations with benefit only emerging in the future, tend to have a relatively slow rate of adoption (Rogers 2003, 233).

Since innovation diffusion is determined by the *perceived* relative advantage, an econometrically computed cost-benefit ratio can only be an estimator for the perceived one. Human perceptions are biased and can deviate from scientific calculations.

- **Compatibility:** Compatibility is "the degree to which the innovation is consistent with the existing values, past experiences and needs of potential adopters." (Rogers 2003, 15). It can be assumed that a society´s acceptance of an innovation is strongly influenced by the innovation´s compatibility, understood in the way defined above.

- **Complexity:** Complexity is "the degree to which an innovation is perceived as difficult to understand and use" (Rogers 2003, 16). A high complexity of an innovation can drive up the transaction costs of its adoption and implementation since uncertainty reduction requires costly research.

- **Trialability:** Trialability is "the degree to which an innovation may be experimented on a limited basis" (Rogers 2003, 16). For innovations that can be tested on a small scale – for example in pilot projects – uncertainty can be reduced and the innovation can be optimized faster than for large-scale and indivisible innovations.

- **Observability:** Observability is "the degree to which an innovation is visible" to other potential adopters (Rogers 2003, 16). Visibility stimulates discussions among peers about the innovation and early adopters can pass on their experience more easily to other potential adopters.

Concluding, the pace of innovation diffusion increases, the more potential adopters perceive the innovation as relatively advantageous, compatible, triable, observable and less complex. According to empirical findings, presented by Rogers (2003, 17), the most important variables on the rate of adoption are the perceived relative advantage and compatibility.

Innovation diffusion processes differ depending on the type and the character of the adopting unit. For instance, the adopting unit can be a single individual, a group of individuals or a formalized organization. For each type of the adopting unit, innovation diffusion faces specific problems. What all innovation diffusion processes have in common, though: The adoption and implementation of an innovation requires a *decision* by an adopting unit.

There are different kinds of innovation decisions (Rogers 2003, 28 f.). For the application of smart grids, most of the relevant decisions can be considered "authority innovation decisions" (Rogers 2003, 28), since in most cases, the decisions to implement new grid-based energy technologies are made by governmental authorities, companies or power utilities. Grid-based electricity supply is a public good which is usually implemented by a central agency. Only to a very limited extend, the adoption of smart grid technologies can be assumed to be optional innovation decisions (decisions of an individual independent from other members of the social system (Rogers 2003, 28)) or collective innovation decisions (consensus among members of a social system (ibid.)). An example of an optional innovation decision can be the voluntary implementation of a smart meter in an individual´s household, a collective innovation decision could be a democratically decided implementation of a mini-grid in a remote village by the village´s general assembly. Adoption decisions for non-grid-based smart energy solutions such as off-grid home systems are decisions, typically made by individuals or households, such that optional or collective adoption decisions are possible. That is, in contrast to main grid implementation, off-grid solutions are usually not adopted due to an authority decision.

Each innovation diffusion research shall include the possibility that the innovation is completely rejected or implemented in a modified way. Modification or *re-invention* is quite typical in diffusion processes, since individuals adapt the innovation to specific framework conditions in their environment (Rogers 2003, 106).

4.2. Innovation diffusion among individuals

This sub chapter takes at close look at the process of adoption decisions which are not embedded in the institutional framework of a formalized organization. The adopting unit is a single person or a rather informal group of individuals. That is, organization-specific influences do not occur. Instead, the adoption decision is rather influenced by the social, cultural and economic environment of the adopting unit.

The innovation decision is a complex process. It takes the decision-making unit from knowing about the innovation, to forming an attitude toward the innovation, to the decision itself, to the implementation and finally to the confirmation of the decision (Rogers 2003, 170). To gain a better understanding about what determines the decision over an innovation, the decision process is divided into separate stages in order to create a mental framework of how the innovation decision works. It can be assumed that under real-world conditions, the different stages are not clearly separated from each. Rather, the stages are connected by fluent transitions. As already hinted before, the chronological stages of the innovation decision are: Knowledge, Persuasion, Decision, Implementation and Confirmation (Rogers 2003, 170).

Figure 3 presents a model of the stages of the innovation decision process. The remainder of this chapter will take a closer look at these stages.

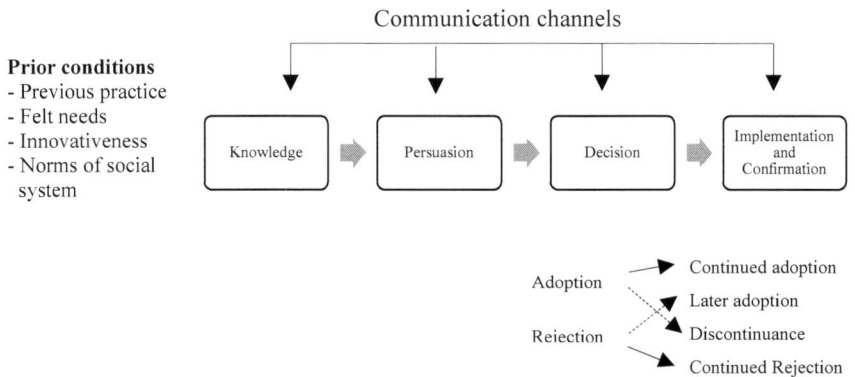

Figure 3: *Model of the innovation decision process.*
Source: Own illustration based on Rogers (2003, 170)

4.2.1. Knowledge

A qualified decision over the use of an innovation requires sufficient knowledge about the new idea. At the knowledge stage of the innovation decision process, three types of knowledge are created.

Awareness knowledge: Awareness knowledge consists of the information that an innovation exists. Rogers (2003, 171) and Hassinger (1959) argue that there are at least two filters on the individual level which an innovation has to pass through, such that the individual knows about an innovation. Firstly, a tendency exists that individuals rather expose themselves to ideas that are in accordance with their specific interests, needs and already existing attitudes. Secondly, an individual does not perceive all innovations it is exposed to. Instead, the individual only perceives those innovations which it finds relevant for its specific situation.

Especially mass media communication channels can spread the information that a certain new technology exists very effectively (Rogers 2003, 173). A mass media channel is "a means of transmitting messages that involve a mass medium" (Rogers 2003, 205). In mass media channels, a new idea can be transported from one sender to very many receivers, such that many receivers create awareness knowledge about this new idea. Examples for mass media communication channels are radio, television and newspapers. The internet is a platform for a combination of mass media and interpersonal communication channels. An interpersonal channel is a "face-to-face exchange between two or more individuals" (Rogers 2003, 205). Using websites or social media, one person can reach many other people with his or her information which is a typical characteristic of mass media communication. On the other hand, the internet also makes interpersonal communication possible, for example in chats and messenger apps.

To generate awareness knowledge about an innovation, especially communication channels that link potential adopters with information sources outside their narrow social system are important (Rogers 2003, 207). Typically, people need information from "outside" to hear about new ideas. These so called, "cosmopolitan channels" (ibid.), may occur as mass media channels but also as interpersonal communication, for example on congresses or trade fares.

How-to-knowledge: How-to-knowledge consists of "information necessary to use an innovation properly" (Rogers 2003, 173). In the case of a smart energy solution, the adopter must understand how to operate this new technology properly to exploit all its benefits. This type of knowledge can be acquired on the job or by additional training. It does not require deeper understanding of the underlying physical or mechanical constructions of the innovation. Knowing how to *use* the technology is sufficient.

Principles knowledge: In contrast to how-to-knowledge, principles knowledge refers to a deeper understanding of the innovation. Principles knowledge is defined as "information dealing with the functional principles underlying how an innovation works" (Rogers 2003, 173). A typical communication channel for principles knowledge is formal education.

If the understanding of the functional principles of an innovation is lacking the introduction and maintenance of an innovation is more difficult and requires additional information-seeking to fill the gaps of principles knowledge. In the case of a smart grid, some basic technological understanding of a smart power management as well as knowledge about information and communication technology are prerequisites for a successful implementation and operation. The importance of principles knowledge once again shows that staff recruitment is an important determinant of an organization´s innovativeness.

4.2.2. Persuasion

At the persuasion stage, individuals form a favorable or unfavorable attitude toward the innovation (Rogers 2003, 174). The necessary condition for an individual to form an attitude toward an innovation is that he or she knows about the innovation and its qualities. Therefore, the persuasion stage is preceded by the knowledge stage. It shall be emphasized that at this stage, it is not about rational reasoning about the qualities, advantages and disadvantages of the innovation but what matters is how the individual *feels* about the innovation. Only if the individual feels positively about the innovation, that is if the individual creates a positive attitude, he or she further bothers about the innovation and the innovation decision process continues. Hence, the type of thinking at the persuasion stage is not cognitive but mostly affective (Rogers 2003, 175).

As already analyzed in chapter 0, certain perceived characteristics determine if the decision-making unit forms a positive or a negative attitude toward the innovation. These characteristics of the innovation are first and foremost: Relative advantage, compatibility, complexity, trialability and observability. The decision-making unit evaluates the quality of the innovation regarding these determinants, takes them all into consideration, weighs them and eventually forms its attitude toward the innovation based on this evaluation.

Looking at communication channels, at the persuasion stage, especially interpersonal communication is important. A positive attitude toward an innovation can be created more easily by close personal communication than via impersonal mass communication channels. Especially positive experience with an innovation from trusted individuals is effective in convincing others to adopt an innovation, too (Rogers 2003, 174 and 205).

4.2.3. Decision

In the light of his or her knowledge and his or her attitude toward the innovation, the individual eventually decides whether or not to make use of the innovation. If this decision is positive, the adoption of the innovation is completed and implementation can start. Formally defined, at the decision stage of the innovation decision process, the individual "engages in activities that lead to a choice to adopt or reject an innovation." (Rogers 2003, 178).

4.2.4. Implementation and confirmation

If the innovation decision process leads to a positive decision, the adoption of the innovation – the decision to make full use of the innovation – is completed and implementation starts. During implementation, the adopting unit puts the innovation into use. While before, the innovation process has merely been a "mental exercise of thinking and deciding" (Rogers 2003, 179), after adoption, the innovation process becomes practical. Like the initiation of the innovation process, implementation consist of different stages which will be described in the following chapters. During the implementation process and the use of the innovation, the adopting unit constantly "seeks reinforcement for the innovation decision already made" (Rogers 2003, 182). At this point, the adopting unit finds itself at the *confirmation stage.* If conflicting feedback occurs at the confirmation stage, the adopting unit might reverse its innovation decision.

4.3. Innovation diffusion in organizations

Since grid-based power supply technology is usually adopted by companies or formalized collaborations of several individuals, the diffusion process in a formal organizational setting is of special interest for the purpose of this study. In this chapter, different forms of formalized collaborations of individuals, such as private companies, governmentally owned energy utilities or authorities are summarized by the term "organization". Precisely defined, an organization is "a stable system of individuals who work together to achieve common goals through a hierarchy of ranks and a division of labor" (Rogers 2003, 404). An energy utility or a formalized collaboration of individuals with the goal to implement power supply infrastructure are typical examples for organizations which meet this definition. The innovativeness of an organization is the degree to which an organization is earlier in adopting new ideas than the other organizations in a social system (Rogers 2003, 475).

What are the determinants of the innovativeness of an organization? Organizations are dynamically efficient, that is successful in developing and implementing innovations, if they are *learning organizations* (Senge 2006). A learning organization encourages advancement through structures and strategies which facilitate and bolster change, innovation and professional development (Dogson 1993, O´Keefe 2002). The core element of a learning organization is a "learning culture" (O´Keefe 2002, 130). A learning culture is a culture that values knowledge and innovation and thus, creates a favorable environment for exploration and experimentation (Hamel, Prahalad 1991). An organization´s culture is built upon the collective organizational memory. Organizations have a cognitive system which is formed by the sum of the organization´s members and which is preserved over time although staff and leadership may change (Fiol, Lyles 1985, Hedberg 1989). Mechanisms which enable the preservation of an organization´s memory are the formation of common values, the history of past events and their interpretation among the organization´s members (O´Keefe 2002, 134).

In an organization that fulfills the prerequisites of a learning organization, change and the regular implementation of new ideas and technologies are self-evident. A learning, open-minded organization can implement innovations much faster, more effectively and with a higher benefit than organizations with a rather conservative culture. Since innovation and change typically come with additional work or even discomfort in the beginning of implementation, a pro-innovation attitude and eagerness to fully understand the innovation helps to reduce resistance against the implementation of an innovation among an organization´s members (O´Keefe 2002, 137). A pro-innovation and learning culture starts at the top of an organization. A management that creates a favorable environment for innovations is typically positively correlated with a higher dynamic efficiency of the whole organization (O´Keefe 2002, 132). Although readiness to learn and a positive attitude toward change among its members is the base of a learning organization, innovativeness on the individual level can only lead to learning on the level of the whole organization if the organization´s decision makers create effective mechanisms and incentive schemes that value, absorb and spread the individual ideas throughout the whole organism of the organization. It is rational for an organization's management to implement such innovation-supportive mechanisms as they enable the organization as a whole to benefit from the innovativeness of its members (Senge 2006, 130).

What sometimes makes genuine reliance on employee initiative and delegation of responsibilities difficult for managers, is that they give away a certain share of their control and power. Bad decisions with far-reaching consequences from employees may result, while the person held responsible for them might still be the manager who delegated the decision. Therefore, each manager faces a trade-off between the benefits and risks of power delegation. The manager's weighing of the risks and benefits determines the level of power sharing, eventually implemented. In his or her consideration, the manager might be tempted to centralize power and implement highly formalized procedures in order not to lose control. Centralization means the degree to which "power and control in a system are concentrated in the hands of a relatively few individuals" (Rogers 2003, 412). Formalization measures how bureaucratic an organization is. It is "the degree to which an organization emphasizes its members' following rules and procedures" (ibid.).

Empirical research and theoretical reasoning clearly indicate that the innovativeness of an organization is negatively correlated with the concentration of power and with its level of bureaucracy (O'Keefe 2002, 137). The problem in a centralized and bureaucratic organization is that the range of ideas taken into consideration by the decision makers is restricted by the low number of strong leaders who dominate the organization and by costly formal procedures. Furthermore, in a strongly centralized and hierarchical organization, problems on the operational level might be neglected as they are not present to the decision makers. Nevertheless, usually the leaders necessarily have to be included into the innovation diffusion process at some point because they typically have the last say in whether or not an innovation shall be implemented. Furthermore, if an organization's leaders actively support an innovation, its implementation can be bolstered significantly (Rogers 2003, 412). It can be concluded, that an innovative company or organization needs a balanced level of power concentration and formalization.

Surprising as it may seem to some people, in many cases, the size of an organization is positively correlated with its innovativeness (Rogers 2003, 405). Reasons might be larger funds for research and development, scale effects and free resources to manage the innovation diffusion process (Rogers 2003, 411). However, it shall be reminded from the preceding reasoning that these benefits of organizational size can only be fully exploited if the growth of the organization does not coincide with an excessive level of concentration of power and formalization.

A further typical finding is that young organizations value creativity and put innovation at the center of the organization's goals. Older, more mature companies tend to be caught in their routine and the implementation of innovations may be more challenging (O'Keefe 2002, 132).

Since an organization's innovativeness is a result of the innovative thinking of its members (Senge 2006, 7), innovativeness demands highly qualified staff. According to Senge (2006), for an organization to be innovative, its staff must be able to think in systems and to include the interconnections of the organization's divisions in their reasoning. That is, limiting the analysis to the own division or the own responsibilities can lead to the neglection of important information and disguise possibilities for improvement. The staff, furthermore, should present a high level of proficiency, the ability to implement effective teamwork procedures and be able to distinguish between important and less important information. To meet these high standards for their staff, learning organizations need high-quality recruitment and a high commitment to capacity building and training. Furthermore, a shared vision of the organization's members which values innovative thinking can increase its dynamic efficiency (Senge 2006, 7-9).

Not only an organization's inner structure and habits determine its innovativeness but also its position and interconnectedness in the social system or the market it operates in. Interconnectedness is the degree to which the organization is linked with other units by interpersonal networks. Being embedded in a diverse and large network, the chance that the organization's members hear about new innovations increases (Rogers 2003, 412).

There are different ways for an organization to be innovative. Either an organization tackles its problems with own effort in research and development to achieve the necessary innovations or an organization scans the environment and its networks for new ideas that might be beneficial. Rogers (2003, 422) calls this second type of corporate innovation policy "opportunistic surveillance". Economically spoken, companies who wait for external innovations benefit from positive externalities. A positive externality is present if at least one exogenous effect with a positive influence exists which is neither controlled nor paid for by the benefitted organization itself (Fritsch 2011, 80 f.).

This external effect can be an innovation, produced by another company or organization. The adopting company benefits from the innovation because in many cases adopting an already existing innovation is cheaper than developing it oneself. Consequently, a company which copies innovations of others acts as a free rider. Free riding behavior and the resulting positive externalities occur if an organization cannot be fully excluded from the use of an external innovation. If the innovator does not possess full property rights of his or her innovation, others can use it, without paying the innovator an adequate price. These opportunistic adopters get some or the whole innovation for free (Rüttgers 2009, 42, Fritsch 2011, 90). The potential to benefit from such positive externalities creates an incentive for organizations to rather follow the strategy of opportunistic surveillance instead of putting effort in own research and development. Economic politics can tackle this lack of incentive for research and development by regulation such as patent legislation which basically assigns property rights to innovators.

The diffusion of innovations in organizations can be illustrated as a process with several steps. After the extensive analysis of the environment which shapes innovation diffusion in organizations, we shall now take a closer look at this diffusion process itself. The process describes how an organization moves from the possibility to adopt an innovation to the final routinized use of the same. Figure 4 presents a graphical scheme for the innovation diffusion process.

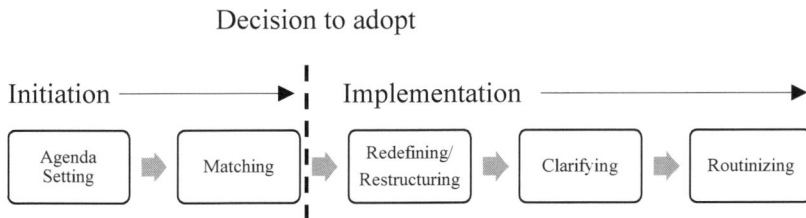

Figure 4: *The innovation process in an organization.*
Source: Own illustration based on Rogers (2003, 421).

Each new step can only start after the completion of the earlier stage. In the remainder of this chapter, the different steps of the process will be described in detail and their interconnections will be outlined. Analyzing the diffusion process, several determinants and prerequisites for innovation diffusion in organizations become evident.

4.3.1. Agenda setting

A necessary condition for an innovation to enter a company or another type of an organization is that the organization perceives a certain need and looks for possibilities to address this need. Precisely, this recognition of a need for some kind of change – called "agenda setting" (Rogers 2003, 421) – is defined as follows: "A general organizational problem is defined that creates a perceived need for an innovation" (Rogers 2003, 422). In here, a need is understood as a "state of dissatisfaction or frustration that occurs when an individual´s desire outweighs the individual´s actualities" (Rogers 2203, 172).

Staying with the example of smart grids in Mozambique, agenda setting might occur in Mozambique´s energy utility EDM in the way that EDM recognizes its distribution and transmission problems and perceives the need to improve its load management in order to reduce transmission losses and blackouts. It should be mentioned that the perceived need can occur due to discomfort with certain processes or existing technologies but it can also be the result of the knowledge about the existence of an innovation. It is possible that an organization perceives a certain need because it hears about an innovation that addresses this need. As March (1981, 568) puts it, "answers often precede questions". The existence of the innovation creates a need which has not been there, before (Rogers 2003, 422 f.).

The argument that sometimes the innovation does not follow the need but the need follows the innovation, is especially relevant for cases when the adoption of an innovation occurs as leapfrogging. Advancements of existing technologies – for example the step from a conventional to a smart grid – do not necessarily only improve the existing technology but also offer possibilities which the older technology did not offer at all. These new qualities might lead additional individuals or organizations to adopt the new technology although they did not adopt the preceding technology. Thus, they leapfrog the older technology and directly start with the new one.

4.3.2. Matching

At the matching stage of the innovation process in organizations, a problem from the organization´s agenda is matched with a certain innovation. Matching describes the planning and design of this process of assigning an innovation to a certain problem (Rogers 2003, 423). Now, the organization knows its problem and how to address it.

Agenda setting and matching form the *initiation* of the innovation process in an organization. "All the information gathering, conceptualizing and planning for the adoption of an innovation, leading up to the decision to adopt" (Rogers 2003, 420) are covered in these two steps.

4.3.3. Decision to adopt

After matching has been completed, a decision whether to implement the innovation has to take place. The decision to make use of an innovation is a core element of the innovation process since it determines if the innovation process continues or – in the case of a rejection of the innovation – never enters the implementation stages. Therefore, the decision separates initiation and implementation in the innovation process.

The decision over the introduction of the innovation is based on the results of the initiation and it connects the stages *matching* with the following stages of the innovation process. The introduction of an innovation in an organization depends on the decision of certain individuals who have the power to execute the decision and spread it throughout the organization. Therefore, the analysis of the innovation decision takes the look away from the general organizational level and shifts toward the individual level. Consequently, the decision process in organizations follows the same steps as extensively described for individual innovation decisions in the preceding chapter, starting at getting to know the innovation across forming an attitude towards it to the adoption decision itself. Eventually, after a positive adoption decision, the steps of implementation and conformation follow (Rogers 2003, 170). It should be mentioned that the decision process is not clearly separated from initiation and implementation but rather overlaps with the last incidents in initiation and the first steps of the following implementation (Rogers 2003, 195).

It has been mentioned before that adoption decisions in organizations are typically authority decisions. That is, although the decision process is the same, the type of innovation decision differs substantially if a decision is made inside or outside a formalized organizational setting. For the typical authority decision in a company, energy utility, governmental institution or any other kind of organization, the choice of rejecting or adopting an innovation is made by one or only a few individuals who possess "power, status, or technical expertise" (Rogers 2003, 29). The organizational framework might influence the outcomes of the different stages in the decision process, such that an individual´s decision to adopt can differ whether it is made inside or outside an organizational setting. For example, financial constraints, legal procedures and accountabilities usually differ between a formalized organization and an independent individual or an informal group of individuals.

4.3.4. Redefining and restructuring

Breaking down innovation implementation in organizations to its single stages, the first one is redefining and restructuring. At this stage, the organization´s staff adapts the innovation to the individual organizational character and its specific problems. In this process, redefining and restructuring of the innovation occurs when "the innovation is re-invented so as to accommodate the organization´s needs and structure more closely, and when the organization´s structure is modified to fit with the innovation." (Rogers 2003, 424). In the case of smart grids in Mozambique, EDM or a private competitor might pursue the re-invention of smart grid solutions in order to meet the specific challenges of the Mozambican context. Smart grid solutions might need re-invention to fit the requirements of small remote villages or a relatively low number of electric devices per household.

What shall be highlighted about the definition of redefining and restructuring above, is that usually not only the organization changes the innovation but also the innovation changes the organization (Rogers 2003, 424). For example, EDM might have to employ new staff or qualify its existing team to develop the skills to handle smart technologies. Furthermore, a cultural change among EDM´s employees and managers might be necessary since digital technologies come with revolutionary changes in the workflow.

4.3.5. Clarifying

Right after implementation and redefining, the new innovation might still cause questions and doubts among some of the staff members. Old routines might be changed by the innovation and some members of the organization might find it difficult to use and understand the innovation. At the stage of clarifying, the innovation gradually enters the different divisions of the organization and its use becomes more intensive. The innovation is gradually used more effectively and consequently, its benefits become more evident to the members of the organization. As a result, the meaning, the sense and the advantages of the innovation become clearer to the organization´s members (Rogers 2003, 427).

4.3.6. Routinizing

In the earlier stages of the innovation process, the innovation was still quite new to the organization and its members. After the initial difficulties have passed, the use of the innovation becomes self-evident and routinizing occurs. Routinizing is defined as the process "when an innovation has become incorporated into the regular activities of the organization and has lost its separate identity" (Rogers 2003, 449). As soon as a routine in the use of the formerly new idea is established, the innovation process is completed. The former innovation is now a natural element of the organization and its workflow.

As this chapter and its preceding analysis have shown, for an innovation process to lead to a successful and routinized implementation of an innovation, many prerequisites – regarding the characteristics of the innovation itself, the characteristics of the adopting unit and its environment – have to be fulfilled. Thus, the principles of the following analysis are provided. The goal will be to analyze whether, to what extend and in which way the prerequisites for a successful innovation process of smart energy solutions are fulfilled in Mozambique. Furthermore, these chapters clearly outlined that large parts of the innovation process are basically information-seeking processes. Thus, it becomes evident, that the quality of information spreading and the corresponding transaction costs are important determinants for innovation diffusion which shall be closely addressed by the following analysis, as well.

5. Drivers and barriers to a smart electrification in Mozambique

5.1. The empirical study

5.1.1. Methodological approach

This study is based on qualitative methods. Qualitative research is appreciated for its depth, concreteness and the understanding of complexity. In contrast, the strengths of quantitative research are a high controllability and the possibility of clear statements about statistical correlations (Heinze 2001, Starr 2014).

Applying qualitative methods, this study is built on a methodology which is well established in social sciences and psychology (Lamnek 2010). Also in economics, the use of qualitative methods is increasing.[5] Qualitative methods are be able to add scientific insights in areas where conventional quantitative research has struggled so far.

Smart energy in Africa has not yet attracted substantial scientific attention such that the analysis of this topic remains scarce. Hence, information about the potential of smart energy applications in the Mozambican electricity sector is found very little and extensive scientific literature analyzing the issue does not exist. Furthermore, the Mozambican government only publishes very few and often outdated data. Due to this lack of quantitative data and specific literature, qualitative methods are promising to fill the gaps.

Another reason to advocate qualitative methods as the main tool for analyzing the Mozambican context of smart energy implementation goes together with the characteristics of quantitative data. Interesting circumstances of economic decisions, political dynamics and cultural specificities vanish to a certain extend if data is quantified (Myers 2013). Due to its open character, qualitative research is considered as a tool that is very sensitive for the specificity of the context of research (Starr 2014). Thus, it

---

5   Bewley (1995, 1999), Lerner, Tirole (2000), Clark, Burgess, Harrison (2000), Desaigues (2001), Chilton/Hutchinson (2003), Böhringer, Löschel (2005), Bird/Higgens/McKay (2010), Valente (2011).

can deliver deeper insights about what exactly might hinder or support smart energy in the Mozambican context.

Human perspectives are an important aspect of this study. Subjective *assessments* about drivers and barriers – especially if held by influential actors – are crucial for the diffusion of innovations: If a sufficiently large share of relevant actors thinks of a variable as a barrier, it will consequently affect decisions as a barrier, regardless if political or economic theory supports this assessment or not. The same applies for potential drivers. Qualitative research is a useful tool to collect information about such subjective assessments (Starr 2014).

Nevertheless, for this study it is regarded a fruitful enhancement to add a quantitative element to the mainly qualitative methods. Specifically, brief questionnaires are issued to a sample of experts for the Mozambican energy sector in order to identify important questions and main topics for the subsequent detailed qualitative interviews. Qualitative research is strong in explaining, quantitative tools such as scale questions are more productive for ranking variables. In this study, the goal is not only to explain the drivers and barriers but also to prioritize them. Thus, further analysis can focus especially on variables which according to the respondents tend to have a strong occurrence in Mozambique. Although a non-randomized and relatively small poll cannot deliver results that can be generalized, it can give an overview of the respondents´ assessments of how strongly they think a driver or barrier occurs in Mozambique.

5.1.2. Basics of the research design

It has already been mentioned that there is very little research experience about the potential of smart energy in the Mozambican electricity sector. Accumulating expert knowledge about the topic of interest with a combination of qualitative interviews and a brief expert poll is considered a promising strategy to fill the gaps. For this method, it is crucial to identify suitable respondents. The point of interest is not the respondent as a person but his or her information. Concentrating knowledge about a certain issue, the expert is thought of being able to deliver valuable information, representative for a larger group of people. This advantage of experts as respondents justifies the concentration of research to a few individuals.

Within this study, a person is considered a suitable respondent, if he or she possesses accessible knowledge that can be used to identify, prioritize and/or explain drivers and barriers to smart grids in Mozambique. Respondents should display sophisticated knowledge about the Mozambican energy sector in general and about the relevant variables that influence political and economic decisions in this sector.

In a narrow sense, experts can be scientific scholars who are relatively independent from the political-economic decision process. Also, economic or political actors can serve as suitable respondents because they usually have a vast experience in the sector of interest and have accumulated specific knowledge. It must be mentioned, though, that actors from the sector under analysis might have more incentives to answer strategically than relatively independent respondents.

To reduce threats to reliability from strategic answers, respondents shall represent the variety of relevant stakeholders (Mayer 2008, 42). That is, if a regulator is interviewed, the regulated company shall be interviewed, too. If a large governmental utility is interviewed, small-scale private competitors shall be interviewed, too. If a sample of respondents consists completely or partly of interest-driven actors, special attention shall be placed on distinguishing factual from subjective information and statements, made on behalf of the respondent´s own interest (Kaiser, Kaminski 2012, 256 ff.).

Since smart energy is quite a new technology, especially to Mozambique, in-depth knowledge about different forms of smart grids and smart off-grid systems is expected to be rather rare among potential respondents in Mozambique. Therefore, the interviews deal with drivers and barriers to smart grids in general without specifying their meaning for individual smart energy *solutions*. Although, technically, the term "smart grid" does not cover all possible smart energy solutions specifically, it is more commonly used than the term "smart energy solutions". To avoid misinterpretation, the questionnaire uses the common term "smart grid", assuming that respondents would interpret this term as a summarizing expression for different kinds of smart energy solutions.

The respondent´s information should be based on actual experiences or education in the area of interest, not on speculations. A high probability to meet these criteria can be found among people who are working in planning, implementation, enforcement, control or development of solutions to problems that are strongly connected to the sector of analysis. Furthermore, people who are involved in decision-making processes that influence the area of interest and people with high-quality access to relevant information are suitable respondents for expert interviews (Meuser, Nagel

1991). It should be assured that the respondent holds the position which makes him or her an expert because he or she is actually qualified and not for other reasons (e.g. party membership, family relations).

For the purpose of this study, respondents were recruited from the following areas: Energy utilities, regulatory agencies, further companies in the Mozambican electricity sector, decision makers in energy politics, donor agencies[6], research and consultancy.[7] That is, the sample of respondents consist of scientific experts and actors from the political-economic process. As already mentioned before, under these circumstances a careful and differentiated evaluation is necessary in order to detect strategic answers from actors which are personally involved in decision making in the power sector.

When experts are found, the next step is to make them bring out the relevant information. For the purpose of this study, it is useful to combine open and closed questions. While the respondent can answer to open questions in own words, closed questions make the respondent choose from given answering-options (Kromrey 2009, 352). Open questions are useful to explore so far unknown and complex topics. They can generate a general understanding and reveal connections between contents by allowing for a detailed and deep analysis. In this study, open questions were used to identify drivers and barriers, to understand how these drivers and barriers affect economic decisions and to reveal the elements that form a driver or a barrier. Additionally, open questions aim to discover specific characteristics like geographical differences of a driver´s or barrier´s impact.

Closed questions can be used to test hypotheses, to clarify or confirm information and to generate numerical data which is open to quantitative evaluation (Mayer 2008, Myers 2013). Answers to closed questions are highly comparable and falsifiable, which is a necessary condition to confirm the validity of research (Popper 1989). In this study, closed questions aim to reveal the following information: Does the sample of respondents regard the variable x in its specific characteristics in the Mozambican context as a driver or a barrier to the implementation of smart grids? To what *extent* does the sample of respondents think that the variable x affects decisions regarding the implementation of smart grids in Mozambique?

---

6   Due to a high level of donor presence in Mozambique´s energy sector, donors are an important actor.

7   A similar selection is used in Ahlborg, Hammar (2014).

The tools in presenting questions to experts are a brief questionnaire to identify key issues and a guide for the semi-structured interviews. While the questionnaire only contains closed questions, the semi-structured interviews focus on open questions. In the interviews, closed questions are only used to specify information, gathered by open interviewing. The complete questionnaire and the interview guide accompanied by a methodological explanation of the tools can be found in annex A and B.

The respondents were consulted in Portuguese, English or German, depending on the respondent´s preferred language. To ensure comparability, a close and adequate translation of the questionnaire and the interview guide was crucial.

5.1.3. Potential drivers and barriers

In order to determine which aspects shall be included in the empirical analysis, potential drivers and barriers have to be derived. These hypotheses about areas where drivers and barriers can occur, form the ground on which the questionnaire and the interview guide are constructed.

How successfully smart energy solutions can be implemented, depends on the answers to the following questions: 1. Can potential users pay for the smart energy solutions under consideration? 2. Are they willing to spend the disposable capital on smart energy? 3. Which quantity and quality of smart energy infrastructure can be implemented under the given budget constraints and the political, economic and social framework conditions? The first two questions relate to the ability and willingness to pay. The third question refers to the environment of smart energy implementation. These overall categories of drivers and barriers – ability to pay, willingness to pay and the political-economic framework conditions – allow for a further differentiation between areas where drivers and barriers to smart energy in Mozambique could potentially occur. These potential sources of drivers and barriers will be presented in the following subchapters.

Finance

It is clear that investments can only be realized if financing is ensured. Therefore, a lack of capital would constitute a serious barrier to smart energy in Mozambique. In contrast to this, a high probability that financially sound investors will put effort in smart energy in Mozambique would push smart grid implementation. Whether or not sufficient capital for smart grid investments in Mozambique can be activated depends on further sub-variables that shall be introduced in the following.

Ability to pay

The ability to pay is determined by the *availability of capital* for the investment under consideration. Availability of capital is broadly specified as the availability of domestic capital on the one hand and foreign capital on the other. A further differentiation is possible between equity capital and borrowed capital. Borrowed capital is money, companies borrow from other institutions, such as banks, while equity capital is the difference between a company´s assets and its borrowed capital (Blanchard, Illing 2009, 657). It has to be considered that a company´s equity capital only enhances the company´s ability to pay if the equity capital is *available* for investments. Only if held as liquid funds, equity capital is immediately available. In contrast, fixed assets are not available for direct financing of new investments (Ricardo 1817, 23). Therefore, the share of market revenues which a company earns as a surplus for future investments is a core determinant for the ability to pay. It is this surplus that determines the level of liquidity of a company in the long run.

The availability of sufficient funds is a necessary condition for the realization of infrastructure investments in general. Since capital is a relatively homogenous production factor, domestic and foreign capital as well as liquid equity capital and borrowed capital can be assumed to be close substitutes from an investor´s point of view. Therefore, investors in Mozambique are probably widely indifferent regarding the source of the capital as long as the capital costs do not differ substantially. What concerns the investors is if capital is available or not.

To address potential shortcomings in the investment environment, policymakers need to know, where exactly capital is short. While the investor might be indifferent where his or her money comes from, the policymaker has to know which specific problems have to be addressed in order to enhance ability to pay for investments.

Willingness to pay

While the availability of capital is only the necessary condition for investments, the existence of a substantial willingness to pay is the sufficient one: Only if investors are *willing* to spend their available capital on smart energy solutions, a smart energy sector can be installed in Mozambique. The willingness to pay depends on the degree to which an innovation like a smart grid meets the felt needs of the consumers. Therefore, the compatibility with the needs of the target group is an important influence of the rate of adoption (Rogers 2003, 246).

In a technical definition, the willingness to pay of consumers is determined by the *reservation price.* The reservation price is the highest price, an investor is willing to pay for a given quantity of infrastructure – here for electrification and smart energy. This definition assumes a fixed quantity of the product at sale and a reservation price that adapts to the quantity (Breidert 2006, 25). However, this relation also holds, if the price is fixed and the quantity is variable: a consumer´s willingness to pay determines which quantity of a commodity the consumer is willing to buy at a given price (Meyer et al. 2007, 226). The foundations of this relation of price and corresponding demand are the consumer´s preferences. The consumer compares different alternatives for spending his or her money based on the utility, the alternatives offer to the consumer (Jehle, Reny 2010).

Aggregated to a whole economy, the willingness to pay for smart energy expresses to which extent all potential investors are willing to spend their disposable capital on smart energy. It follows that the willingness to pay for smart energy is a core determinant for the quantity of smart solutions, eventually installed. Therefore, the empirical analysis will ask the question to which extend potential investors are willing to spend their available capital on smart energy. As the previous chapter has shown, there are different kinds of investors with different attributes. Probably not only the ability but also the willingness to pay differs among these different kinds of investors. Therefore, the empirical analysis shall cover these potential differences.

Electricity market

Even if high amounts of capital are activated, further influences for the eventually installed quality and capacity of an investment remain. One important factor are the characteristics of the market which the investment under consideration is supposed to enter. If the market works well and no serious distortions exist, no considerable barriers to the implementation of an investment arise. If, however, market failures – market power, high transaction costs, information disparities – or distorting regulative interventions into the market exist, the outcome of the investment will probably be biased and unsatisfying. There is a high risk that under such adverse conditions, a large part of the disposable capital is not used efficiently, less capacity than possible is installed or that the commodity comes with a lower quality than possible. The determinants of the Mozambican market´s performance and potential sources of market failures are introduced in the following.

Market power

The level of competition is a core determinant for the efficiency of every market. Economic theory states that a perfect and fully competitive market serves best to generate welfare. Whenever competition is comprised despite functioning market processes, the market outcome is biased and welfare losses may result (Meyer et al. 2007, Jehle, Reny 2010).

To understand the impact of market power on the welfare of a society, it is fruitful to compare a situation with market power to a theoretical model with ideal-typical assumptions where welfare is maximized. In a perfect market, the welfare of the market participants is maximized if the price of the product equals the marginal costs of the producers. (Meyer 2007 et al., Jehle, Reny 2010). The marginal costs are defined as the increase of the total production costs, induced by the production of an additional unit (Mankiw, Taylor 2008, 307). In an environment of complete competition, the function of the marginal costs equals the supply function of the producers (Meyer 2007 et al., Jehle Reny 2010). *Figure 5* illustrates the maximization of welfare in a perfect market equilibrium. The market equilibrium is achieved when demand equals supply. If demand equals supply an equilibrium price emerges, that equals the marginal costs for the resulting quantity of the product x^*. The sum of the area between the demand curve and the price (consumer surplus) and the area between the price p^* and the marginal costs (producer surplus) is the total welfare of the market participants. It reaches its maximum level in the market equilibrium where the price p^* equals the marginal costs $mc(x^*)$.[8]

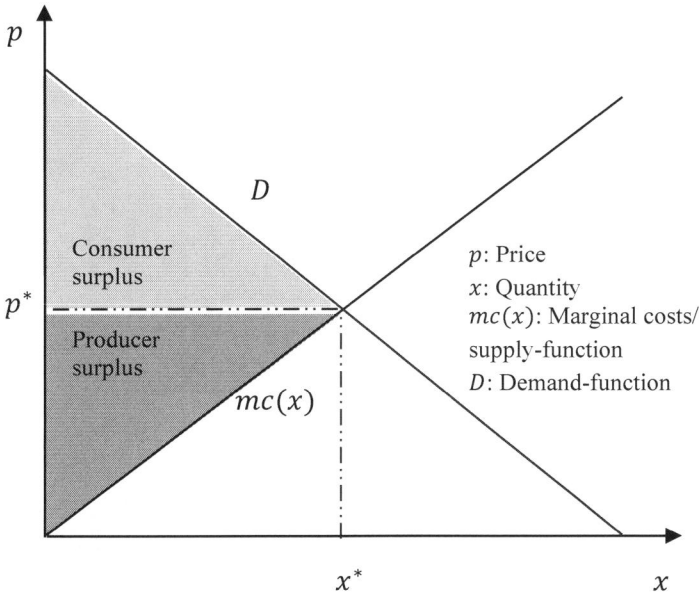

Figure 5: *Equilibrium in a market with perfect competition.*
Source: Own illustration based on Meyer et al. (2007).

In a market that is characterized by supply-side market power, the welfare-maximizing equalization of price and marginal costs is not achieved. The following reasoning illustrates this problem in the case of a monopolistic market. The monopoly knows the demand function $x^D(p)$ with x^D being the demanded quantity of the product depending on the price p. The demand function can be resolved into $p(x^D)$. The monopoly can choose the quantity of the product x^S which it supplies. The condition for a market equilibrium is

---

8   In order to avoid extensive economic reasoning, the theoretical foundations of welfare maximization are postulated without further explanation. For an extensive derivation of the postulated relations see Meyer et al. 2007 and Jehle, Reny 2010.

$$x^S = x^D \tag{1}$$

So, for the equilibrium, we can replace x^D in the resolved demand function $p(x^D)$ such that

$$p = p(x^S) \tag{2}$$

Relation (2) is the price-demand function of the monopoly. The revenue R of the monopoly is its sold quantity of the product x^S multiplied with its price. Hence, we can write the revenue function as

$$R = px^S = p(x^S) \times x^S \tag{3}$$

The monopoly´s profit π is the difference between its revenue R and its costs C.

$$\pi(x^S) = R(x^S) - C(x^S) \tag{4}$$

The monopoly maximizes its profit. To acquire the necessary condition for a profit maximum we form the first derivation of the profit function π and set it equal to zero.

$$\frac{d\pi}{dx^S} = \frac{dR}{dx^S} - \frac{dC}{dx^S} = 0 \tag{5}$$

For the profit-maximizing solution, it follows that

$$\frac{dR}{dx^S} = \frac{dC}{dx^S} \tag{6}$$

The expression dR/dx^S is the marginal revenue of the monopoly and dC/dx^S is the monopoly´s marginal costs. As (6) shows, in the profit maximum of the monopoly, marginal revenue equals the marginal costs. We gain the marginal revenue of the monopoly by forming the first derivation of the revenue function in relation (3)

$$\frac{dR}{dx^S} = p + x^S \times \frac{dp}{dx^S} \tag{7}$$

We can reorganize (7) such that we gain an expression called Amoroso-Robinson-Relation (Robinson 1932)

$$\frac{dR}{dx^S} = p + x^S \times \frac{dp}{dx^S} = p \times \left(1 + \frac{x^S}{p} \times \frac{dp}{dx^S}\right) = p \times \left(1 + \frac{1}{\varepsilon(x)}\right) \qquad (8)$$

In relation (8), $\varepsilon(x)$ is the price elasticity of demand $\varepsilon = dx/dp \times p/x$. It shows the relative change in demand that follows from a relative change in the price (Meyer 2007, 74). In a fully competitive market, the price elasticity of demand approaches infinity, since even a very small price increase above the equilibrium price by one single supplier would drive away all its former customers to other suppliers and the demand for the single, more expensive supplier drops to zero. For normal commodities, like electricity can be assumed to be, it is reasonable to assume that the demand decreases as the price increases. Therefore, according to the definition given above, the expression $\varepsilon(x)$ forms negative values for normal commodities.

As we have already shown in (5) and (6), in the profit maximizing situation, marginal revenue equals the marginal costs. So, we can replace the marginal revenue in (8) by the marginal costs.

$$\frac{dC}{dx^S} = p \times \left(1 + \frac{1}{\varepsilon(x)}\right) \qquad (9)$$

Given the typical attributes for $\varepsilon(x)$ for a normal commodity in a non-competitive market, $\varepsilon(x)$ can be assumed to be negative and finite in the relevant domain. Furthermore, it is only rational for the monopoly to sell its product if the price elasticity of demand $\varepsilon(x)$ is larger than one in its absolute value: If $\varepsilon(x)$ is smaller than one in its absolute value, according to equation (9), the price p would have to be below the marginal costs $\frac{dC}{dx^S}$ or even negative (in the case of $\varepsilon(x)$ being negative and smaller than one). A price below the supplier's marginal costs or a negative price would cause losses and consequently, the supplier would cease the production.

Under these reasonable conditions for the value of $\varepsilon(x)$ (negative, finite and larger than one in its absolute value) it follows that $p > dC/dx^S$ in the monopoly's profit maximizing allocation. This relation unambiguously shows: in a monopolistic market, the price is higher than the marginal costs (Jehle, Reny 2010). As the formal reasoning showed, the reason is, that in the monopolistic market, the market equilibrium follows the profit maximization condition of the monopoly instead of resulting from the competition of many suppliers. As we have already concluded earlier, in the welfare maximizing market equilibrium, the price equals the marginal costs, though. That is, in the case of market power, the price is too high. The higher price results in a lower demand for the product and therefore in an inefficiently low quantity of the commodity. Thus, market power results in welfare losses – the higher the market power, the higher the welfare losses (Jehle, Reny 2010).

The preceding reasoning has important implications for the implementation of smart grids in Mozambique. As the economic theory has shown, an unregulated, monopolistic energy utility would have strong incentives to sell less power than would be sold in a welfare-maximizing market in order to narrow supply and thus drive up the price. The less power is sold by the supplier, the less need is there for grid infrastructure. An increase of market volume to an efficient level would mean that more power would have to be sold and new grids – potentially smart ones – would have to be implemented. An increase of the market volume could either be induced by a regulatory agency or by market entry of new competitors. For example, new suppliers could connect formerly unconnected households and increase the capacity of their power grid. Thus, market volume would be driven up and new grid infrastructure would be necessary.

Besides its effects on price and quantity of the product, a lack of competition also has consequences for the *dynamic efficiency* of a market. In a *dynamically efficient* market, welfare will be increased over time by the implementation of new technologies. To achieve a high level of innovativeness, a market has to produce effective incentives for the introduction of new technologies (Fritsch 2014). However, if only one supplier or a small group of suppliers dominates the market, there is a risk of biased incentives. Over- or underinvestment in research and development or in the implementation of new technologies is a likely result. For example, a strong and irreversible monopolistic energy utility comes with a significant probability to be inefficiently reluctant to new technologies such as smart grids because the monopoly does not face the risk that other, more innovative competitors could challenge its dominating position in the market. Furthermore, a monopoly has incentives to take efforts to crowd-out competitors and to prevent new companies from entering the market (Meyer at al. 2007). This misuse of market power can impede small-scale investments in decentral grid-solutions and other bottom up initiatives (Ahlborg, Hammar 2014).

Another problem of market power arises regarding the quality of the product – in this study the supply of energy. In the absence of competitors, the monopoly has an incentive to increase its profit by decreasing the quality of its product. In the case of an energy market, this adverse incentive could result in a low number of connected households, regular black-outs or load shedding with negative effects on the consumer´s welfare.

The preceding reasoning showed that the level of market power in the Mozambican power market has a high potential to influence the quantity and quality of smart energy implementation in Mozambique and the price for energy services. In order to analyze the level of competition in the Mozambican energy market and to reveal the consequences for the implementation of smart energy, the issue of market power shall be included in the empirical analysis.

Tariffs

The level of electricity prices is a core determinant of the power companies´ revenues. Only if the revenue is high enough for some surplus to remain, the company has funds to finance investments, for example in smart grids. Therefore, too low tariffs would eventually narrow the possibilities for smart energy investments in Mozambique. Especially maximum tariffs constitute a severe barrier to market entry of private companies.

Not only the level, but also the structure of a tariff regime shapes the performance of the power sector and the expected benefits of a smart grid. It can be distinguished between fixed and flexible tariff regimes. Fixed tariffs are always at the same level, no matter what time of the day or what level of electricity demand is present. In contrast to this, flexible tariffs charge more when power supply runs short and less when a lot of energy is available in the system. Flexible tariffs create incentives to use less power in times of high scarcity and drives consumers with a low willingness-to-pay out of the market when power is short (Schreiber et al. 2015). In effect, energy-supply can be made safer and more efficient, since high peak-load-prices (Steiner 1957) are avoided.

What are the individual dynamics that make this price-induced demand response possible? The following formal reasoning illustrates in an example of a flexible tariff structure with a lower off-peak price p_o and a higher peak price p_p, how a representative rational consumer adapts his or her consumption patterns.

We assume that the consumer aims to reduce his or her electricity costs C under the condition that his or her utility from electricity consumption does not fall below a certain level of minimal utility $\bar{u}$. This setting reflects a common consumption pattern with an electricity consumption that follows a usual daily routine consisting of activities such as heating, cooking, lighting and using technical devices that produce a constant flow of utility.

To model the consumer's utility, we apply a common Cobb-Douglas utility function $u = x_o^\alpha x_p^\beta$ (Cobb, Douglas 1928). The variable x_o represents power use in off-peak periods and x_p stands for the power use in peak periods. Since power in off-peak periods and power in peak periods are nearly perfect substitutes – the consumer cannot distinguish between their quality – the consumer has no incentive to adapt his or her consumption patterns to the current scarcity if there are no price differences between peak power and off-peak power. It is an important property of a Cobb-Douglas utility function that it generally allows for the substitution of the consumed goods x_o and x_p without a change in utility (Cobb, Douglas 1928, Douglas 1976). This property of a Cobb-Douglas utility function reflects the assumption that to a certain degree it is possible for the consumer to substitute electricity use in peak periods with electricity use in off-peak periods and vice versa. For example, it is reasonable to assume that a representative consumer does not suffer a significant decrease of utility if the consumer's phone is charged a few hours earlier or later than usually if the consumer does not need the phone urgently (for example at night when the consumer sleeps). Consequently, phone charging in periods of low demand (off-peak) is a close substitute of charging in peak periods. However, there are limits to the substitution of electricity in peak and off-peak periods. Although the quality of electricity is the same in peak and off-peak periods, the consumer might want to use electricity in peak periods – even if it is more expensive – because this consumption pattern accords to the consumer's daily routine. For example, cooking dinner or watching the evening news on TV is fixed to a certain time of the day. If this time of the day coincides with peak power demand, the consumer will probably still use electricity unless the peak price becomes prohibitively high. Taking all things into consideration, substitution between peak and off-peak electricity demand is generally possible without a loss in utility but only to a limited extend – like in the Cobb-Douglas utility function.

If, however, the price for off-peak power is different than the price for peak power, things look different. In the case of differentiated prices, the consumer has an incentive to distribute his or her demand over peak and off-peak periods such that the consumer receives the electricity he or she needs at the lowest possible costs.

To which amount the representative consumer uses power in off-peak and peak periods if the prices differ, is a typical economic decision problem that can be solved by a few mathematical operations. The representative consumer's optimization problem is fully described by:

$$\min_{x_o, x_p} C = x_o p_o + x_p p_p \quad \text{s.t.} \quad x_o^\alpha x_p^\beta \geq \bar{u} \tag{10}$$

That is, the consumer aims to minimize the costs (quantity of x_o and x_p, each multiplied with their specific prices) under the condition that his or her utility, represented by the Cobb-Douglas function, does not fall below the level of $\bar{u}$.

To find a solution to this optimization problem, we form the corresponding Lagrange-function.

$$\mathcal{L} = x_o p_o + x_p p_p - \lambda(x_o^\alpha x_p^\beta - \bar{u}) \tag{11}$$

After some mathematical operations, a Lagrange-function delivers the necessary conditions for the solution of the optimization problem.[9] The Lagrange function in (11) yields the necessary condition for a minimum-cost solution as follows:

$$\frac{x_p}{x_o} = \frac{\beta}{\alpha} \frac{p_o}{p_p} \tag{12}$$

Now, it can be analyzed what happens, if tariffs are not flexible but fixed, that is if the peak and the off-peak price are equal: If $p_o = p_p$, the equation in (12) reduces to

$$\frac{x_p}{x_o} = \frac{\beta}{\alpha} \tag{13}$$

In (13), the relation of off-peak and peak electricity consumption does not depend on the price relation any more, but only on the consumer's preferences. That is, in a fixed tariff regime, there is no adaptation of the consumer's demand to price variations. If, however, prices differ between dif-

---

9   What a Lagrange-function is, how it is formed and how it is used is extensively described in Jehle, Reny (2010) on page 579 ff.

ferent periods, the consumer adapts his or her electricity consumption. For example, if the peak price p_p in (12) increases, the proportion of peak consumption x_p in x_p/x_o decreases.

Breaking it down to the basics, this finding indicates that in a flexible tariff regime, peak demand decreases in relation to off-peak demand if prices are higher in peak periods than in off-peak periods. In effect, such a flexible tariff regime induces a relatively lower demand if electricity is short and a higher demand if it is vastly available. In effect, the consumer´s electricity consumption is smoothened across peak and off-peak periods.

As the formal operations have shown, flexible tariffs create incentives for demand response by passing scarcity signals on to the consumers. This contribution of the consumers has the potential for cost savings on the consumers´ side, addressing the low ability to pay in Mozambican households and at the same time leads to a lower utilization of the very expensive peak capacity. The harmonization of demand and supply facilitates the integration of intermittent renewable energies into the energy system.

There are different approaches for a flexible tariff regime. Tariffs can be higher in times of the day when, according to prior experience, demand for energy is very high. Another possibility is to adjust tariffs continuously in a real-time process to the present levels of demand and supply. For both designs of a flexible tariff regime, but especially for real-time adjustment, a smart energy system is needed to collect and process the necessary data about consumption and production patterns. Only high-quality time series data about demand and supply enable to determine the efficient price level for each period. Especially if tariffs are to be adjusted in a real-time process to the current scarcity of power, sophisticated data analysis and enhanced metering technologies are needed.

Hence, a combination of a smart energy system and flexible pricing creates the possibility for a precise demand side management. In effect, not only the producers contribute to the load management but also the consumers, motivated by scarcity-adjusted tariffs (Schreiber et al. 2015). Thus, the loads of power in the electricity system can be balanced more effectively, the risk of blackouts and load shedding are alleviated and the need of additional generation capacities to tackle peaks of demand is reduced.

It has clearly been stated that the price level and the degree to which the tariff structure is flexible shape the chances and the expected benefits of smart energy implementation. Both aspects shall therefore be part of the empirical analysis.

Transaction costs

Transaction-costs arise from using a social-economic system as a measure of coordination. Transactions on imperfect markets require collecting information about other parties, resource-intensive negotiating and enforcing contracts. During these processes, the necessary co-ordination and organization comes with costs (Coase 1937, 390-392).

In the case of smart grids, the implementing organization – in here mainly EDM – has to analyze carefully which quantity, quality and design of smart solutions fit the organization´s specific needs and requirements. Otherwise the producers and suppliers of smart technologies could exploit the lack of information on EDM´s side. It has to be considered that the producers of a technology are typically more informed about its quality and production costs than the buyer – an asymmetric distribution of information is present. In this setting, the seller has an incentive to act opportunistically, that is, to use the information deficit of the transaction partner for its self-interest, even if doing so harms the transaction partner (Williamson 1985, 47).

For example, the seller could put a higher price on the transaction or reduce the quality of the traded technology while the buyer is unable to observe the quality differences entirely. It becomes evident that in a setting of asymmetrically distributed information between seller and buyer, extensive research and controlling is required on the buyer´s side to reduce the possibilities of opportunistic behavior on the seller´s side.

The buyer´s risk to be harmed by opportunistic behavior of a seller who exploits his or her information advantage is even aggravated if the transaction deals with investments of certain characteristics. Especially technologies or products which´s quality is difficult to observe, come with high incentives for the abuse of informational advantages. When important characteristics are only known by one party but hidden to the other one, the better informed party has a strong incentive to exploit the less informed one. Therefore, the less informed party will invest a large effort to reveal the hidden characteristics which means high transaction costs. In contrast, products which can be tested on a small scale at low costs and with an easily observable quality are correlated with rather low transaction costs. This is one of the reasons why innovations tend to be adopted more quickly if they come with a high level of trialability and observability (Rogers 2003, 221).

The installation of power grid infrastructure typically causes high transaction costs. Grid infrastructure is a costly and at the same time specific infrastructure. "Specific" means in this context, that grid infrastructure cannot be used in other production processes but power transmission. That is, if the investment fails in some way – for example if the resulting benefits are much smaller than expected – there are no alternative uses that could compensate for the losses. Therefore, especially for specific investments with high implementation costs and an unobservable performance, careful analysis of the costs and benefits is required before the transaction takes place in order to avoid bad surprises afterwards (Williamson 1989, 145). These additional requirements for information gathering are probably the main reason for high transaction costs in the case of costly and highly specific investments like power grid infrastructure and especially smart grids.

What does more, especially new or very complex technologies bring about a high need for information and testing. Especially regarding new and complex technologies, their quality and benefits are difficult to observe. As chapter 0 has already shown, the innovation-decision-process is basically an information-seeking process. A high complexity combined with little experience can increase information-seeking activities significantly and thus reduce the rate of adoption for the product under consideration (Rogers 2003, 221). Therefore, higher transaction costs can rather be expected for the implementation of new and complex technologies than for such technologies which can be operated without much prior training or for which a lot of experience already exists (Hartwig 2004).

In the case of legal insecurities or shortcomings in the legislation, transactions costs arise due to the need for legal consulting. In countries, where the rule of law is comprised, investments might fail, for example as a result of unexpected expropriation. Aiming to avoid failing investments due to a difficult legal environment, investors typically need additional legal consulting that drives up transaction costs.

Legal insecurity can occur frequently for new technologies as the legal framework might not yet have been adjusted entirely to the new situation. For example, regulative measures that were very effective for analog processes might need adjustment when these processes are object to a profound digitalization. An illustration for this argument is the discussion about smart meters. To fully exploit the advantages of a smart grid, consumers have to implement smart meters (Dada 2014). Hence, a legal framework that is compatible with the implementation of smart meters can bolster intelligent grids while a regulation, fitted to conventional meters, would impede the digitalization of the power system. Further sources of high transaction costs can be inefficient planning and complicated administrative procedures. Only if administration works efficiently while bureaucracy is kept at the unavoidable level and necessary documents are issued quickly, economic activity can develop without major delays or distortions.

Moderate transaction costs occur within practically every transaction on markets. However, transaction costs can distort the market significantly if they reach a critical level. If transaction costs are prohibitively high, the transaction might even not take place at all. In contrast to this, an economic environment that offers efficient and effective means for information procurement, well working legal institutions, legal security and qualified staff, helps to keep transaction costs low and can be a driver for the implementation of new technologies. Therefore, the question shall be included in the empirical analysis, to what extent transaction costs can be expected to occur in the implementation process of smart grids and how they effect this process.

Infrastructure

As the analysis of innovation diffusion in chapter 4 has shown, compatibility of the innovation with the environment of the adopting unit is an important determinant for a successful adoption of the innovation. Technical applicability is one component of compatibility. Smart grids could be applied in different ways: Either the existing grid system is upgraded with smart components or a completely new grid system is installed. In the first case, the smart grid components have to be compatible with the existing grid. A lack of interoperability (Welsch et al. 2013) of conventional and smart grid can cause problems. In the case of completely new smart power lines, isolated mini grids and off-grid solutions, compatibility challenges with an existing grid are avoided. However, also a new grid or off-grid infrastructure can face compatibility problems. Questions like the following arise: Is the infrastructure compatible with typical building techniques

(Ahlborg, Hammar 2014)? Is it compatible with the skills of power sector staff? Is it compatible with the existing needs of society?

Since the Mozambican electricity grid does not reach many parts of the country, the problems with existing grids are assumed to be less relevant than in countries completely covered with an existing power grid. For the many white spots on Mozambique´s electric power map, leapfrogging can be a promising option. Furthermore, if the existing grid is old and has to be replaced, the new grid can be a smart one. The implementation of smart technologies in Mozambique could be facilitated by an already existing and quickly expanding infrastructure of mobile information- and communication technology (ICT). Like many African countries, Mozambique leapfrogged to mobile ICT-solutions, driven by the quickly expanding cell-phone-market (INE 2014, World Bank 2014). Replacing or upgrading still acceptably working infrastructure causes opportunity costs, as there is still the option to keep the old grid running. Due to its specificity, the costs of the existing grid are sunk and therefore, incentives to replace it are low.

Besides technical applicability, another important factor for the implementation of energy infrastructure is the ratio of its benefits toward its costs. As the analysis of innovation diffusion in chapter 0 has shown, the perceived relative advantage of an innovation – the ratio of the expected benefits and the costs of adoption (Rogers 2003, 223) – is an important determinant of the rate of adoption. Empirical research by diffusion scholars has even shown that the perceived relative advantage is one of the strongest predictors of the rate of adoption (Rogers 2003, 221). Hence, relatively high expected benefits in relation to relatively low expected costs of smart energy in Mozambique would clearly create a driver to smart grid implementation. Vice versa, a disadvantageous cost-benefit ratio would indicate a low perceived relative advantage and therefore constitute a barrier.

The long-term costs of a technology are not only determined by the costs of adoption and implementation but also by its costs, arising from maintenance and operation. Maintenance and operation of smart grids require distinct knowledge and capabilities. A lack of specifically trained power sector staff might be a barrier to smart grid implementation (Welsch et al. 2013). Even if the necessary knowledge is available, a loose culture of maintenance can be a constraint for a high long-term performance of smart grid (Bugaje 2006).

Governance

The economic and political development of a country is strongly shaped by the performance of its political institutions and political actors. Stable, democratically legitimate and efficient institutions, transparent political decision-making and the rule of law are important factors for an attractive business environment. Therefore, good governance – or put more precisely: development-oriented governance (Kevenhörster 2014, 15) – is a prerequisite for the development of a strong, efficient and smart energy sector.

"Governance" refers to how political decision-making takes place as well as how policies are formulated and implemented. Therefore, governance does not only describe the observable actions of the government but also political processes and the participation of the private sector and civil society. It consists of the quality and structure of the institutions but also of informal settings, such as the organizational culture, the integrity of the staff or path dependencies (Kersting et al. 2009, 161).

Political actors may be committed to overall well-being and progress but it is also possible that they opportunistically focus on their own benefit at the expense of their fellow citizens. Governance can be qualified as good and development-oriented if it acts in accordance with the preferences of the citizens and achieves output-legitimacy by effectively delivering public services and common welfare (Kersting et al. 2009, 9). Furthermore, good governance is characterized by processes and institutions which show an economical use of resources and a responsible use of power to provide public goods for an inclusive and sustainable development. Organizations and institutions shall furthermore act transparently, accountably, independently of political influence and resistantly to corruption (Klemp, Poeschke 2005, 20).[10] A strong and alert civil-society that is actively involved in decision making can control the government, bring up new ideas and thus foster development-oriented governance.

A core determinant for the quality of governance is a country´s regulatory framework. Trade barriers to power supply components, legally restricted access to technologies (Painuly 2001, 79), missing data-

---

10 For a more detailed definition of the term "good governance" see Klemp, Poeschke (2005), who developed a set of benchmarks as a framework for good governance.

transparency and a strong regulation of the energy sector can distort the economy. Looking at smart grids in particular, a poor regulatory framework can make smart grids less beneficial and lower the possibilities of successful implementation. To avoid expensive tailored solutions, the regulatory and commercial framework affecting power supply technologies, should be harmonized – on national and international level. This also implies, that technological standards for components and especially for relatively new technologies such as smart grids are necessary (Welsch et al. 2013).

For the Mozambican context, there are some severe threats to good governance which can prove harmful for the energy sector's development. One of these threats is the ongoing political and occasionally violent conflict between the ruling party FRELIMO, the largest opposition party RENAMO and their military wings. The conflict itself will be described in more detail in chapter 0. Already at this point, though, it shall be emphasized that severe political and especially violent conflicts can lead into a situation where economic and political development is basically impossible. A violent escalation of diverging political interests takes a country into a conflict trap: As political priorities shift from development towards the fights, development impulses are blocked at the origin (Collier 2008, 33).

Another strong indicator of bad governance is corruption. Corruption is mainly a result of rent-seeking behavior. Rent-seeking refers to pursuing the increase of the own share of the existing wealth without creating new wealth. That is, rent seeking aims to create an own benefit through a redistribution of given funds at the expense of the other members of society instead of improving one's own situation through productive behavior. Since rent-seeking comes with effort and time, resources are used for nonproductive actions. The resulting inefficient allocation of resources leads to a lower overall wealth such that justice problems occur, since rent-seekers acquire wealth which does not correspond with a productive contribution (Tullock 1967).

The more governmental agents show openness to bribes in exchange for privileges, the more they encourage rent-seeking behavior. If companies or individuals feel that using their resources for bribes gets them quicker and easier to their goal than investing into productive use, rent-seeking can be expected to increase. Consequently, the struggle for re-distribution is bolstered at the expense of productivity (Olson 1982).

Experiences in many countries show that economic policies, mainly focused on the extraction of natural resources should not be considered as development-oriented governance. Perhaps in contrast to a first intuition, rich fossil resources can counteract economic and political development (Sachs, Warner 1999). One important economic reason is that the increasing export of natural resources leads to an appreciation of the national currency. As a result of this appreciation, the competitiveness of other commodities from the country deteriorates. This phenomenon is also referred to as the "Dutch disease" (The Economist 1977, 82). Further reasons why fossil resources can counteract an inclusive and sustainable development are an increased vulnerability to corruption, stronger short term-orientation due to high rents in the present, dependence on resources and crowding out of other manufacturing and services (Venables 2016).

If Mozambican decision-makers can easily be seduced by rent-seeking opportunities, the country is presumably under a high risk to be affected by such resource traps. Offshore gas reserves are estimated at 277 trillion cubic feet, estimated offshore coal reserves sum up to more than 20 trillion tons. Exploitation and exploration of offshore-resources are recently increasing and deposits might be even much larger than currently assumed. Plus, proven onshore natural gas reserves of approximately 3.5 trillion cubic feet add up to the resource base (Cipriano et al. 2015). This large availability of fossil resources and correspondingly the high expected benefits create strong incentives for a run on the resources which can be expected to come with several temptations for rent-seeking and even corruption.

This chapter summarized several potential political and regulatory threats and chances for smart energy in Mozambique. Due to these various potential impacts on electrification in general and smart energy implementation in particular, the issues of governance and regulation should be an important part of the analysis of drivers and barriers to smart energy solutions in Mozambique.

International Cooperation

Economic cooperation with like-minded countries can foster smart energy implementation and lead to an improvement of the Mozambican electricity sector in general. International networks create cosmopolitan communication channels which bolster the diffusion of innovations, as the analysis of diffusion processes in chapter 0 has already shown. Members of international networks benefit from an exchange of experience and get in touch with new ideas and technologies. Best practice examples from one member of the network can be copied by other members. Furthermore, a joint implementation of programs enables the pooling of funds and can produce the output more efficiently due to a better division of labor and responsibilities.

Especially for recipient countries of large amounts of development aid, development cooperation is an important aspect of international cooperation. Mozambique´s economy and government strongly depend on development aid (see ch. 0). If smart energy implementation is among the donors´ priorities, development aid can become an important driver for the digitalization and the improvement of the Mozambican power sector. The more effectively donors perform, the greater the support.

An effective development cooperation brings about a *sustainable* improvement that diffuses through the recipient´s economy and society and is *not limited to the direct target group* of specific development projects. There are many factors that determine whether or not development cooperation is eventually effective. Some of the core criteria are the following (Kevenhörster 2009, 82 f.):

- Productive orientation: The donor generates a productive outcome for the target group.
- System orientation: Innovation leads to diffusion processes that go beyond the closer target group and lead to an improvement of the whole system
- Innovation-orientation: The donor possesses sufficient innovative power such that the donor can react adequately to changing framework conditions.
- Coordination and coherence among donors: Donors coordinate their actions and ensure the coherence of their policies and goals. A lack of coordination among donors or a lack of coherence among donors´ policies or goals can increase transaction costs and harm outcomes (see also: Ahlborg, Hammar 2014).
- Relevance and impact for development: The donor´s activities are focused on projects with the largest benefit for the recipient.
- Efficiency: The donor uses its resources economically.
- Sustainability: Positive effects remain after the donor´s involvement is ended

- Constant evaluation and improvement: The donor evaluates the projects and improves its work where shortcomings are revealed.
- Recipient orientation: The donor chooses projects which are coherent with the needs and goals of the recipient. Sustainable success in development cooperation can only be achieved if the target group accepts the donor´s goals already at the beginning of the project and still after the project has finished (see also: Kevenhörster 2014, 16)
- Ability to absorb of recipient: The recipient´s economic, social and political system is able to absorb the development aid in an effective and efficient way such that a positive output is produced. Especially bad governance or a lack of economic power can lead into a situation where donor support has no additional use and cannot be absorbed. In a recipient country with a very low ability to absorb, development aid faces quickly diminishing returns (see also: Kevenhörster 2009, 90).

For a long time, development cooperation has been criticized for a too strong input orientation. That is, donors are accused of focusing too much on how much money or effort they contribute without considering sufficiently which actual impact their projects created for the recipient (Kevenhörster 2014, 11). Furthermore, in many cases, scientific evaluation reveals that development projects receive a positive evaluation for the direct effect on the target group but seem to fail to produce positive effects on economic progress and poverty relief on the large scale. This phenomenon is briefly summarized by the term *micro-macro-paradox* (Faust, Leiderer 2008, 129).

It is possible, that the micro-macro-paradox is simply a result of an attribution gap. It is very difficult to attribute developments on the large scale, like economic growth, to single development projects. Research that tries to do so comes with large methodological problems especially due to bias from omitted or unobservable variables. Hence, the attribution gap is the domain in which no logical, clearly provable effects can be drawn from a development activity since further potential factors are not controlled by development cooperation. Macro-scale developments can therefore not be fully attributed to efforts on the micro level due to an unmanageable amount of potential influences (Kevenhörster 2014, 21)

However, there can also be more serious reasons for the observation that development aid might not lead to improvements on the large scale. Development aid can proof as a distortion of the recipient´s economy or even hinder economic development in the long run. Donor money can

79

hinder self-generated initiative, create dependencies and distort econo-mies, especially if donors´ priorities and approaches are not steady (Djankov et al. 2008). Additionally, aid-inflows bring about strong incen-tives for rent-seeking behavior. In effect, development aid can support the alimentation of clientelistic structures and corruption. Thus, the govern-ance of the recipient country can be negatively affected by development aid (Kevenhörster 2014, 16).

Berg-Schlosser and *Kersting* (1996) name political clientelism as an important reason of an inefficient allocation of financial resources, includ-ing development aid, in African countries. The authors describe the rela-tionship between the population and its political representatives in African countries to a large extent as an asymmetric system of exchange in which votes and political support are rewarded with actual or promised material advantages (Berg-Schlosser, Kersting 1996, 103). Thus, political decision-making is seized by clientelistic interests which overweigh other motives for political decisions like political programs, ideology, economic situa-tion, milieu or confession (Berg-Schlosser, Kersting 1996, 104). Conse-quently, in many cases, political parties in Africa are not built on a com-mon ideology or economic class but on the materialistic interest of a dif-fuse and politically heterogenous group of stakeholders. In such an asym-metric system of mutual exchange of interests and loyalties, the temptation is high to misuse political influence in order to secure power by advantag-ing only the "own people".

The inflow of development aid can stabilize this clientelistic exchange structure because donor money opens new potentials and new temptations to benefit the own voter base at the expense of the public interest. Chances are high that a clientelistic allocation contradicts a fair and productive use of aid resources as clientelistic allocation is often not the most productive one. For instance, if staff for development projects is not hired based on qualification but political loyalty, development is impeded.

Furthermore, the inflow of foreign currencies through development aid can lead to an adverse terms-of-trade effect for the recipient´s economy: Financial development aid appreciates the national currency of the recipi-ent, as inflows lead to a higher demand for the receiving country´s curren-cy. Thus, exports from the recipient´s country become more expensive and the country´s economic development is impeded as it was already de-scribed for the occurrence of resource traps (Kevenhörster 2014, 16).

All things considered, despite certain risks, development aid brings about large chances for the development of the recipient country – and specifically for the Mozambican power sector – if it is effective, coherent and supportive of good governance structures. Nevertheless, the risks of development aid have to be taken into consideration.

Acceptance

To what extent and how quickly a smart grid infrastructure spreads through Mozambique, depends on how attractive it appears to investors and suppliers and how well it is accepted by consumers and other relevant stakeholders in society. The more accepted an innovation is, the more rapid its rate of adoption is expected to be (Rogers 2003, 227). Also in the long run, that is during routinizing and confirmation, smart energy implementation can only be sustainable if it is aligned with the specific demands of the adopting units. It has to be considered that perceptions of innovations shape public acceptance, not objective indicators (Rogers 2003, 432). Therefore, the level of acceptance strongly depends on how relevant actors communicate and promote the new idea and how they include the public.

The example of smart meters illustrates quite well how acceptance problems can block the diffusion process of an innovation. Precise harmonization of electricity demand and supply requires a high market penetration of smart meters to monitor consumption profiles (Schreiber et al. 2015). However, costumers might be reluctant to purchase costly smart meters and an obligatory rollout has the potential to raise strong acceptance problems. The necessary collection of personal data and consumption patterns can be expected to raise opposition against smart grids from relevant stakeholders. If the emphasis of data privacy is very present in a society its violation can lead to politically articulated protest against the collection, storage and communication of personal information (Benett 1992). Especially at an early stage of the introduction of digital technologies, anxiety to new technologies can be another important barrier to adoption (Igbaria et al. 1994). What is more, each power grid – be it smart or conventional – requires large infrastructure investments. The corresponding impact on landscape, nature and local communities can cause acceptance problems. In the worst case, infrastructure investments can make resettlements necessary such that strong opposition from affected communities can be expected.

Since the largest part of smart grid implementation is normally undertaken by energy suppliers, we deal with innovation decisions in organizations. Therefore, not only the society´s but also the implementing organization´s staff´s acceptance of the innovation has to be considered. Diffusion theory and empirical analysis of diffusion processes in organizations state that the sustainability of an innovation depends on whether or not the staff accepts the innovation. One important determinant of the staff´s acceptance is, how well the employees are included in the innovation process. The degree of staff participation in designing, discussing and implementing the innovation is an important influence of innovation implementation. Hence, a lack of acceptance and participation among the adopting organization´s staff can constitute a barrier to smart grid implementation from inside the adopting organization (Green 1986).

It has become evident that acceptance and stakeholder participation are important determinants for smart energy diffusion. Therefore, the following analysis of drivers and barriers shall include these aspects.

5.1.4. Application of the questionnaire

The categories for potential drivers and barriers, derived in the preceding chapter, are joined in a questionnaire to be used for the empirical analysis. As it has already been mentioned in chapter 0, the results of the questionnaire come as numerical data which facilitate the prioritization of drivers and barriers. *Table 1* presents an overview of variables, included in the questionnaire. The complete questionnaire and its methodological background can be found in annex A.

The questionnaire can be applied separately from the qualitative interviews. Thus, the questionnaire can introduce additional respondents´ views into the analysis. A large share of the actors and experts from the Mozambican energy sector is located in the capital Maputo as most of the federal authorities, research entities, companies and consultancies are based there. However, an extensive study about Mozambique´s power sector should not only be limited to expertise from the Maputo region. The questionnaire allows for the inclusion of further respondents, for example local EDM-staff from remote provinces. Therefore, the variety of experiences and points of views among the respondents is increased by adding the expert poll to the mainly qualitative methodology of this study.

Table 1: *Questionnaire and corresponding variable names.*

Question (In Mozambique, what is the occurrence of...)	Variable
Available capital for investments in general?	Availability
Willingness to spend available capital on smart grid investments?	Willingness
Competition in the electricity market?	Competition
Transaction costs, induced by smart grid implementation?	Transaction costs
Incentives, induced by power tariffs, to implement smart grid solutions?	Tariffs
Smart grid-supportive governance?	Governance
Smart grid-supportive regulations?	Regulation
Support of smart grid-implementation by donor involvement?	Donors
International cooperation regarding the implementation of new technologies?	International
Technical applicability of smart grid-infrastructure in Mozambique?	Applicability
Expected quality of grid management (maintenance and operation) in Mozambique?	Maintenance
Stakeholder acceptance of a smart grid infrastructure in Mozambique?	Acceptance

Source: Own table

The empirical analysis of drivers and barriers was conducted in summer 2016. In total, 23 experts and actors from the energy sector responded to the questionnaire[11], of which

---

11   A complete list of the experts is attached in appendix C.

- Seven are scientists at the universities of Maputo, Chimoio (central Mozambique) or the Geneva Graduate Institute of International and Development studies,
- Five are EDM-staff,
- Four are consultants from development cooperation agencies (of which two ex-patriots and two locals),
- Two are leading officials at the Mozambican Ministry of Mineral Resources and Energy (one head of department, one leading official of the ministry´s provincial delegation in the Province of Manica (central Mozambique)),
- Two are entrepreneurs (one director of a Mozambican renewables company and one member of board of an international smart energy company),
- One is the director of the German Chamber of Commerce in Maputo,
- One is a former official of the German Embassy in Maputo,
- One is a representative of FUNAE.

With 23 purposely chosen respondents, the sample does not qualify for the application of advanced quantitative evaluation methods. Limited resources, logistic challenges and a small number of energy experts in Mozambique limit the sample. Scientific entities, dealing with the Mozambican power sector are rare, the number of companies and organizations in the Mozambican power sector is low. The respondents had to be chosen purposely since responding to the specific questions requires expert knowledge. Therefore, the data is not eligible for statistical methods which require random sampling.

Nevertheless, the results from this poll bring about important additional value to this study. The prior goal of the expert poll is not to produce generalized statements which infer from the sample to a larger population but to take a closer look at the statements of the group of respondents. First and foremost, the results from the questionnaire should show as how important the sample of respondents assesses the impact of potential drivers and barriers.

Most of the questionnaires were submitted to the experts by email. Seven out of the 23 experts had to answer to the questionnaire on the phone. Reasons are the severe security problems at the time of research (see analysis of the violent conflicts in chapter 5.6.2). Possibly, it poses a threat to reliability if some respondents fill in the questionnaire themselves while others are asked on the phone. It is preferable to make the respondent fill in the questionnaire by him- or herself in order to avoid any influence from the researcher. However, it would have been a very narrow picture if respondents from other areas than around Maputo had been excluded due to challenging political and technological framework conditions. Points of views from people, living in rural areas or communities with different cultural or economic backgrounds are important sources of information, as well. Nevertheless, it shall be stressed that the questionnaire is merely supposed to supplement the core tool, the qualitative interviews, by facilitating the identification of priority determinants of smart energy implementation.

5.1.5. Application of the qualitative interviews

While the expert poll is seen as a useful tool, especially to prioritize broadly formulated drivers and barriers, the qualitative interviews are expected to contribute better to understanding, explaining and subdividing drivers and barriers.

To structure the interviews and to increase comparability, an interview guide is developed. The categories under examination in the interviews are based on the potential drivers and barriers, derived in chapter 5.1.3. Generally, the questions in the interviews are derived from the same overall categories like in the expert poll: ability to pay, willingness to pay and the economic-political framework conditions, shaping smart energy investments. However, the qualitative questioning is much more detailed and focuses more on the origins and effects of the drivers and barriers. The complete interview guide and extensive methodological explanations can be found in annex B.

A total of nine extensive expert interviews with decision-makers from different areas of the electricity sector was realized. Specifically, the following respondents were interviewed:

- Manager of grid and transport, EDM, Directorate of Grid and Transport, Department of Power Lines
- Manager of grid management and maintenance, EDM, Directorate of Grid and Transport, Department of Grid Management,
- Manager of projects and financing, EDM, Directorate of Electrification and Projects,

- In-house consultant of FUNAE, especially in charge of off-grid electrification and mini-hydropower plants,
- Head of Department of Renewable Energies, Ministry of Mineral Resources and Energy of the Republic of Mozambique,
- Director of MOZITAL, medium-sized Mozambican company in the field of renewable energies, founded in 2006, since than broadly 2000 installations in broadly 120 villages,
- Consultant of the Deutsche Gesellschaft für Internationale Zusammenarbeit mbH (GIZ), expert for electrification, especially renewables and off-grid solutions,
- Professor at Geneva Graduate Institute, Graduate Institute of International and Development Studies,
- Professor at Eduardo Mondlane University, Maputo, Chair for Physics of the Renewable Energies. Also, this respondent is the chief executive officer of a small-scale company, selling and installing mini grid and off-grid installations based on renewable energy. Additionally, the respondent is the current president of the general assembly of the Mozambican association of renewable energies.[12]

The interviews lasted between 45 and 75 minutes with seven interviews conducted personally and two via video call (due to logistic reasons). The interviews were ended when the conversation did not appear to reveal any new relevant information. No significant disadvantage of the video calls was noticed. The quality of the conversation was comparable with the personal interviews. Table 2 summarizes the basic research design, derived in this chapter.

---

12   A list of all experts is attached in annex C.

Table 2: *Basic research design.*

	Identify drivers and barriers	Prioritize drivers and barriers	Understand origin and effect of drivers and barriers
Main tool	Interview guide, questionnaire	Questionnaire	Interview guide
Target group	Experts	Experts	Experts
Type of questions	Open and closed	Closed	Open

Source: Own table.

5.1.6. Data evaluation

For the scale questions, each variable is classified as a driver or barrier based on the respondents´ assessments. After identifying drivers and barriers, they are prioritized. The evaluation includes an assessment of the results´ validity. Reliability is the necessary condition of validity. A high level of agreement among respondents is an indicator for a reliable method, since it shows that the method delivers similar results each time it is applied (Dorussen et al. 2005). A high level of agreement among experts is expressed by a relatively small variation of the answers. That is, measures of spread represent the level of agreement among experts. In this study, the standard deviation of the results is computed in order to analyze the spread of the answers.[13]

Yet, validity requires not only that research is reliable, but also that it measures the real drivers and barriers to smart grids in Mozambique. To determine the level of validity, qualitative, numerical and theoretically derived results were cross-checked whether they are coherent with each other or not. Thus, the risk of systematic errors is reduced. If a certain empiri-

---

13   Due to characteristics of the data (e.g. existence of non-numerical data ("don´t know-answers")), other measures of inter-rater-agreement like Kendalls coefficient of concordance (Kendall, Babington Smith 1939, Kendall 1948) do not deliver meaningful results.

cal result contradicts widely accepted aspects of scientific theory or existing empirical data, a sign of a lack of validity is present unless a sound explanation for this deviation can be found (Carus, Ogilvie 2009, Kevenhörster 2014, 26).

For the evaluation of the interviews, the goal is to find patterns such as common opinions or assessments of the experts (Mayer 2008, Starr 2014). Interviews are taped and transcribed in the form of a result protocol to ensure a detailed and efficient evaluation of the results.[14] After transcription, results are ordered into categories in order to find patterns, common statements and contradictions. The organization of the results is accompanied by an analysis of the expert´s arguments´ inner logic. Special focus rests on the identification of drivers and barriers to smart electrification, their relevance, their origins and effects and to the way they influence smart grid investments.

To avoid an influence of strategic, interest-driven answers which bias the result, the evaluation should pay special attention to the following questions (Kaiser, Kaminski 2012, 256 ff.):

1. Which statements can be considered as relatively objective information?
2. Which statements are led by personal interests of the respondent or of the entity, the respondent represents?
3. Which statements are the respondent´s subjective opinion?

Thus, subjective or strategic statements shall be identified and analyzed. This is of special importance for the statements of respondents who are actively involved as actors in the Mozambican power sector. Actors might answer relatively objectively to most of the questions while they have a special personal interest in a certain question and answer strategically. Revealing subjective or strategic statements enables the concentration of the analysis to the reliable statements. While doing so, it shall be kept in mind, though, that this research does not directly lead to political decisions. This reduces the risk of strategic answering by respondents.

---

14 Protocols of the interviews are attached in annex D.

5.1.7. Methodological challenges

In expert interviews and expert polls, the sample cannot be drawn randomly. Experts have to be chosen purposely for the respondents to display high levels of relevant expertise and variety to cover different points of view (Starr 2014). Additionally, the sample size is small in comparison to quantitative research which limits the representativeness of research. However, representativeness for the overall population is not of first importance in expert interviews. Expert interviews and expert polls rather ask questions such as: Are experts competent? Do they offer new insights? Are they exemplary for the different points of views? Is the sample representative of the specific population of experts (Flick 1999, Myers, Newman 2007, Mayer 2008)?

In qualitative research, enlarging a sample brings about the risk to interview a respondent who does not possess the necessary expertise. Hence, the researcher has to determine the optimal size and diversity of the sample (Bewley 2002). No further interviews are to be conducted if they are unlikely to yield any additional insights (Myers 2013). Therefore, sample sizes in qualitative research differ from less than five cases to a few hundred (Starr 2014). However, the specific strength of qualitative research to reveal detailed insights and the assumed higher probability of experts to give correct answers outweigh the small sample size for the purpose of this study.

For the questionnaire, which can be used independently from the semi-structured interviews, the sample size is larger. Nevertheless, the complexity of the questions in the questionnaire is high and therefore, expert knowledge is required. Additionally, the necessary limitation of respondents to experts on the topic of interest makes a randomized and large sample impossible for the questionnaires.

A small sample size poses a potential threat to internal and external validity of the results. Aside from sample size, further threats to validity are a lack of objectiveness on the researcher´s side and dishonest answers by the respondent (Bewley 2002, Lamnek 2010). Qualitative research relies on the performance and the assessments of the researcher. In order to achieve a high level of objectivity, the researcher should force him- or herself to leave personal values aside. The possibility of ideological influences or political interests which could bias the answers, shall be considered in the evaluation. It was offered to each respondent to participate anonymously, in order to avoid incentives to answer strategically. In total, none of the respondents in the interviews and three of the participants in the poll asked to do so (see Annex C). The mentioned threats to validity are addressed in the remainder of this paper by accurate cross-checking of quantitative data, qualitative results political and economic theory.

5.2. Main findings

The empirical study delivered robust and insightful results which help to identify, prioritize and understand the drivers and barriers to smart energy implementation in Mozambique. The bar chart in figure 6 gives a first overview of the results from the questionnaire. Presented are the mean values (arithmetic mean) of the variables. The graph clearly shows that the experts´ evaluation differs substantially from variable to variable, some being rather regarded as barriers, some as drivers. A negative mean indicates a barrier because it represents a rather harmful influence on smart grid implementation or infrastructure investments in general. Furthermore, although the variable transaction costs produces a positive mean, it has to be considered a barrier, too, since high transaction costs mean a challenging environment for business. The remaining variables with positive means indicate drivers since for these variables, a positive mean is a sign for a rather favorable environment for power sector investments and smart grids.

According to the mean values the power *tariffs*, the level of *competition* in the power sector and the *availability* of capital form a cluster of relatively strong barriers. Less severe but still clearly ranked as barriers are the regulative framework *(regulation)*, the *willingness* to pay and *transaction costs*. With a mean value already close to zero but still negative, *maintenance* and *governance* follow as variables that shall be ranked as rather neutral or slightly adverse for smart energy implementation.

Changing the view to the drivers, the relevant stakeholders´ *acceptance* of smart grid investments in Mozambique sticks out as the variable with the highest mean value. As further drivers follow *donor* engagement and *international* networks. Technical *applicability* of smart grids still produces a positive mean which is quite close to zero, though.

Once again, it shall be mentioned that these results only represent the assessments of a sample of purposely chosen experts. In order to infer from these results to the Mozambican context in general, additional reasoning and cross checking with further empirical and theoretical findings is needed and will be given in the proceedings.

Further statistical evaluation can help to determine whether ambiguous variables like *applicability* shall be considered as a relevant influence to smart energy implementation in Mozambique. In this sense, after this first impression of the variables, the remainder of this chapter will take an ever closer look at the statistical characteristics of the results of the empirical study in order to distinguish and prioritize the drivers and barriers.

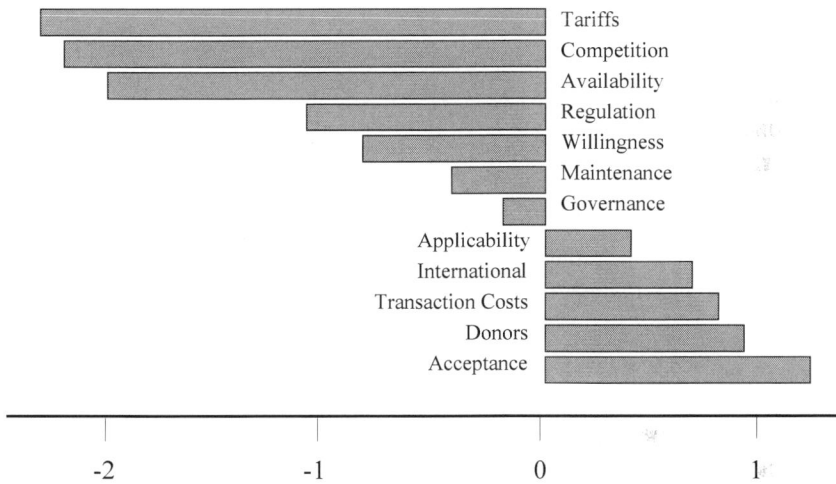

Figure 6: *Mean values of drivers and barriers.*
Source: Own illustration.

Table 3 presents a profound analysis of the empirical data. Mean and median as measures of fit indicate the classification as driving or impeding factors. The standard deviation represents the spread of the answers.

Like already indicated in the bar chart, based on the results of the statistical evaluation shown in the table, the variables *availability (of capital), competition* and *tariffs* were clearly classified as barriers by the respondents. Their average values are quite negative, considering that the scale only reaches until minus three. Therefore, it can be inferred, that the sample of experts considers the variables *availability, competition* and *tariffs* as major barriers among all revealed barriers to smart grid implementation

in Mozambique. In addition, the standard deviations of these three variables are the three lowest under consideration, ranging from 0.85 to 0.96.
This finding is an indication of a relatively high level of agreement among
the experts, showing little variation in their assessments. According to the
assessments of the experts, it can be assumed that the Mozambican economy is characterized by a lack of available capital for investments in general, that competition in the electricity market is very low and power tariffs fail to produce incentives to implement smart grid solutions.

Table 3: *Classification of variables.*

variable	n	mean	median	sd	classification
Availability of capital	19	-1.95	-2	0.91	Barrier
Willingness to pay	16	-0.81	-2	1.72	Barrier
Competition	21	-2.14	-2	0.96	Barrier
Transaction costs	13	0.77	1	1.17	Barrier
Tariffs	20	-2.25	-2	0.85	Barrier
Governance	21	-0.19	0	1.78	NA
Regulation	16	-1.06	-2	1.94	Barrier
International cooperation	17	0.65	1	1.00	Driver
Donors	17	0.88	1	1.54	Driver
Applicability	16	0.375	0.5	1.67	NA
Maintenance	19	-0.42	0	1.84	NA
Acceptance	17	1.18	1	1.42	Driver

Source: Own table.

Further variables, classified as barriers are *willingness (to pay)*, *transaction costs* and *regulation*, because their average values indicate an impeding effect. That is, the respondents´ assessments imply that in Mozambique, the willingness to spend available capital on smart grid investments
is small, transaction costs, induced by smart grid implementation are high
and the regulatory framework lacks smart grid-supportive arrangements.
However, all of the three variables are characterized by a relatively large
variation in the answers, representing somewhat ambiguous expert opinions. While the variable *transaction costs* still produces a moderate standard deviation of 1.17, the standard deviations of *willingness (to pay)* and
regulation reach 1.72 and 1.94.

In the experts´ judgements, a high level of *acceptance* among stake-holders is the strongest driver for the implementation of a smart energy infrastructure. It is the only driver, that reaches a mean value larger than one. It shall be mentioned, though, that its standard deviation (1.42) is relatively high which indicates some ambiguity in the answers.

Further variables, classified as drivers are *donors* and *international cooperation*. Both variables have positive averages. Experts tend to expect a positive effect of donor involvement on smart energy implementation and perceive a positive effect of international cooperation regarding the implementation of new technologies. It should be mentioned, though, that both variables´ averages are smaller than or equal to one which can be an indication for a relatively small impact on smart energy implementation. Looking at the variation in answers, the relatively high standard deviation of *donors* (1.54) indicates rather ambiguous expert opinions. In contrast, *international (cooperation´s)* standard deviation (1.00) is moderate.

For *governance, applicability* and *maintenance*, averages are zero or close to it in combination with relatively high standard deviations. Therefore, neither of these variables is classified as a driver or barrier. That is, according to the average expert opinion, the levels of expected quality of grid management (maintenance and operation), technical applicability of smart grid infrastructure and smart grid-supportive governance are neither extraordinarily high nor extraordinarily low in Mozambique. However, it shall be mentioned, that the standard deviations of the three variables *governance, applicability* and *maintenance* range from 1.67 to 1.84, showing a relatively large spread. Expert assessment on these variables is ambiguous. Whereas some experts regard these variables´ level of occurrence as high, indicating a driver, another notable part thinks that these variables come at low levels, indicating a barrier.

As a first conclusion, the empirical results from the questionnaire indicate the following:

- A lack of availability of capital in general, missing competition in the electricity market and low incentives, induced by electricity tariffs are considered major barriers to smart grid implementation in Mozambique by the sample of respondents,
- Further variables classified as barriers are: A lack of willingness to pay for smart grids, transaction costs, coming with smart grid implementation and a poor regulatory framework,

- The engagement of donors, international cooperation on the implementation of new technologies and a relatively high level of acceptance of smart grid infrastructure are classified as drivers,
- The variables *governance, applicability (of smart grid infrastructure)*
  and *maintenance* could neither be classified as drivers nor barriers.

A further categorization is not pursued since it would not deliver robust statistical differences and is therefore of no informative value for this study.

Having derived first indications for identifying and prioritizing possible drivers and barriers to smart grid implementation in Mozambique, we have not yet gained any insights into the mechanisms, how exactly the drivers and barriers affect economic decisions regarding the implementation of smart energy solutions. Questions like the following are raised: Why is a certain variable a driver or a barrier? What exactly is the problem with power tariffs in Mozambique or what are the reasons for the lack of capital? In order to understand origins and effects of the drivers and barriers, we need to take a closer look at the qualitative interviews. This tool allows in-depth analysis of the different variables and thus contributes to a better understanding of the topic. The following pages present a first overview of the qualitative results.

Table 4 presents the barriers[15], table 5 the drivers[16], mentioned by the respondents in the interviews. The findings contain additional indications of what exactly impedes or supports smart grid implementation in Mozambique. Furthermore, the frequency, a certain driver or barrier is mentioned by the respondents, can be an indicator of its relevance.

---

15 Sources indicated in table 4: 1 = EDM, 2 = FUNAE, 3 = Ministry of Mineral Re-
    sources and Energy, 4 = Entrepreneur, 5 = Donor (consultant), 6 = Scientist de-
    velopment studies, 7 = Scientist renewable energy. Number of interviews with
    each organization: i − iii. B+ and B++ mean that the barrier was mentioned in
    two or three interviews with the same organization. The symbol $\sum$ refers to the
    total number of interviews in which the barrier was mentioned by a respondent.
16 Sources indicated in table 5: 1 = EDM, 2 = FUNAE, 3 = Ministry of Mineral Re-
    sources and Energy, 4 = Entrepreneur, 5 = Donor (consultant), 6 = Scientist de-
    velopment studies, 7 = Scientist renewable energy. Number of interviews with
    each organization: *i − iii*. D+ and D++ mean that the driver was mentioned in
    two or three interviews with the same organization. The symbol $\sum$ refers to the
    total number of interviews in which the driver was mentioned by a respondent.

Table 4: *Barriers mentioned by respondents.*

Identified barrier	Source							Σ
	1	2	3	4	5	6	7	
	iii	*i*	*i*	*i*	*i*	*i*	*i*	
Finance								
Lack of capital in general	B+	B	B			B		5
Lack of domestic capital	B				B	B	B	4
Low willingness to pay		B			B	B		3
Short-term orientation					B	B		2
Electricity market								
High level of market power	B+	B	B	B	B		B	7
Low revenues in electricity market		B	B	B				3
Low power tariffs	B++	B	B	B			B	7
Lack of flexibility in tariff structure	B	B			B		B	4
Transaction costs	B+			B	B	B		5
Economic environment								
Poor infrastructure	B+							2
Lack of qualified staff	B+	B	B		B			5
Low density of population		B	B	B	B		B	5
Lack of initiative		B						1
Environmental impact		B						1
Infrastructure								
Indivisibilities of grid components		B	B	B	B		B	5
Lack of equipment and tools	B+							2
Lack of maintenance	B	B	B		B		B	5
Missing real-time information	B+							2
Economic performance								
Higher costs of smart grid		B		B		B	B	4
Governance								
Political tensions, violent conflict		B	B		B	B	B	5
Poor performance of institutions		B		B	B			3
Centralized decision-making	B	B				B		3
Corruption				B	B			2
Regulatory framework	B+	B		B	B		B	6
Unclear responsibilities							B	1
Lack of ICT-knowledge		B		B				2
Mssing knowledge energy polictics		B			B			2

Source: Own table.

Table 5: *Drivers mentioned by respondents.*

Identified driver	Source 1	2	3	4	5	6	7	Σ
	iii	i	i	i	i	i	i	
Finance								
Availability of foreign capital in general		D				D		2
Availability of donor capital	D+	D	D		D			5
High willingness to pay among donors			D		D		D	3
Economic environment								
Economic growth	D					D		2
Industrialization	D		D		D			3
Increasing agricultural production					D			1
Growth of energy demand					D	D		2
Increasing private sector engagement		D	D					2
Abundance of mobile internet network	D	D	D					3
Potential for renewables	D	D	D	D	D			5
Connection of public facilities	D		D					2
Potential of digitalization								
Potential to reduce transaction costs	D+				D			3
Potential to improve maintenance	D+							2
Increasing use of ICT			D					1
Already existing smart energy projects	D	D	D					3
Potential to improve decentral solutions	D	D	D				D	4
Potential to improve energy supply	D+			D				3
Leapfrogging due to new technologies		D						1
Economic performance								
High benefit of electrification in general	D++	D	D	D	D	D		8
High benefits of smart solutions	D++		D			D		5
Potential of cost reduction for utilities		D		D				2
Leapfrogging due to cost effectiveness						D		1
Reduction of transmission losses	D	D						2
Governance								
Government´s electrification goals	D		D					2
Development cooperation	D	D	D	D	D			5
Acceptance								
of infrastructure projects	D+	D	D	D	D	D		7
of smart energy solutions	D+	D						3

Source: Own table.

A variable, mentioned both as a driver and a barrier, is classified according to its relative frequency. Thus, it is considered a driver if brought up more times as a driver and as a barrier, respectively. If there is diverging assessment about a certain variable, it is mentioned and discussed in the following chapters.

In line with the quantitative evaluation above, a lack of capital in general and of domestic capital in particular was among the barriers, frequently mentioned in the interviews. On the other hand, table 5 shows, that the availability of foreign capital, in here especially capital of donor agencies, was considered a driver by a relevant part of the experts. According to this finding, qualitative research reveals that the variable *availability (of capital)* has to be looked at in a differentiated way: Even though capital availability in general is regarded a major bottleneck to smart energy implementation in Mozambique, the decomposition of this variable shows that smart energy investments relying on foreign or donor capital can expect a higher probability of sufficient funding.

A similar picture emerges with willingness to pay for smart grid solutions which was also classified as a barrier in the quantitative analysis. While three of the experts claim that a low willingness to pay is generally a barrier, the representatives of the Ministry of Mineral Resources and Energy, the GIZ and Eduardo Mondlane University agree that specifically among donors, there is a high willingness to pay for smart grid solutions.

It is reasonable to expect that the overall shortage of capital in Mozambique and the low willingness to pay are caused to a significant extent by the political and economic environment in the country. For example, political tensions and violent conflicts between government and rebels, a factor, frequently mentioned as a barrier, probably have a directly impeding impact on smart grid investments but also an indirect effect by further tightening capital availability. The same holds for the barrier "poor performance of public institutions", the high prevalence of corruption and the poor regulatory framework. All these variables can hinder smart energy investments directly and furthermore scare away potential investors. A distinct short-term orientation among potential investors, brought up by two of the experts, can be another explanation for a low willingness to pay for smart grids because it takes some time for the benefits of a smart energy system to outweigh the high initial investment costs.

As barriers from the area of governance are mentioned: centralized administration, unclear responsibilities, missing qualification regarding information and communication technologies and a lack of knowledge about energy politics. Nevertheless, the variable *governance* could not be classi-fied as a barrier in the quantitative analysis. One reason might be, that the Mozambican government´s ambitious electrification goals (see chapter 2) are regarded as a driver to the implementation of smart grids. This opinion was shared by both EDM´s staff and the representative of the Ministry of Mineral Resources and Energy. Furthermore, official development coop-eration – which also involves national institutions and decision-makers – was mentioned five times as a driver.

Among the economic developments in Mozambique, there are factors that raise optimism among the experts: Strong economic growth, industri-alization and increasing agricultural production, accompanied by a growth of energy demand, can reduce concerns of potential investors and are seen as drivers by some of the experts. The gross domestic product in Mozam-bique is projected to increase by 4.1% in 2018 and 4.7% in 2019 (Caldeira 2018 b). The GDP also grew in the years before with about 4% in 2017 and 2016 (INE 2016, Caldeira 2016) and more than 7% in each of the four years before (UN 2016 a). According to the experts´ statements, further drivers concerning the economic environment are: increasing private sec-tor engagement, the economic potentials of renewable energies, the pro-ceeding grid connection of public facilities such as health centers and the vast potentials of digitalization. Experts state that digitalization brings about potentials to reduce transaction costs, to improve off-grid solutions and to enhance maintenance. Thus, digitalization might be able to address some of the barriers such as high transaction costs (mentioned five times and also classified a barrier in the quantitative part) and missing real-time information about grid performance which was clearly one of EDM´s staff´s major concerns when it comes to maintenance and operation of the grid.

Digitalization is claimed to have a positive impact on energy supply in general. Some experts also mention that first smart energy projects have already been implemented in Mozambique. These pilot projects can drive the implementation of smart grids on a larger scale. An important loca-tional factor, especially for smart energy solutions, is a large geographical coverage of high-quality information and communication technology. Ex-perts attest that Mozambique can score with a good and quickly increasing abundance of mobile internet. The representative of the Ministry of Min-eral Resources and Energy furthermore thinks that the development and availability of new technologies opens potentials for leapfrogging because whenever a new grid is implemented in a place where no transmission network has existed before, investors can directly use the new and smart solutions.

Mentioned seven times, the high acceptance of infrastructure projects in Mozambique is one of the most important drivers. Also for smart grid projects specifically, three experts expect high levels of acceptance. These statements match the quantitative evaluation that also derived the variable *acceptance* as one of the drivers.

On the other hand, central economic factors are given bad marks by the experts. Besides a poor infrastructure and a lack of qualified staff, a further barrier is a lack of initiative among economic actors. FUNAE´s respondent claimed that in Mozambique, the readiness to translate ideas into action is very low. Instead, he claims, a large part of the population rather waits for decisions and activities put in place by others, like foreign investors or donor agencies. It can be expected that these negative properties of Mozambique as an investment location are also reasons for the capital shortness, discussed earlier and manifested in a low availability of capital and willingness to pay. A barrier to smart grid implementation is seen in the environmental impact caused by the necessary infrastructure investments.

Taking a closer look at the Mozambican electricity market, the strongest barriers in the qualitative results concern power tariffs and the competition in the power market, just like in the quantitative evaluation. The barriers "high level of market power" and "low tariffs" are both mentioned by seven of the nine experts, making them the two most frequently mentioned barriers. Experts state that EDM is a typical governmentally protected monopoly that concentrates nearly all the market power in its hands. Furthermore, they claim that power tariffs – which are set by the government – are too low. This goes together with another identified barrier, namely the low revenues in the electricity market. Another shortcoming the experts identified in the tariff structure. They claim that the tariff structure contains too little flexibility, such as time-adjusted tariffs. These findings are a specification of the quantitative results, which only told us, that the variable *tariffs* ("incentives, induced by power tariffs, to implement smart grid solutions") came with a low rating in the expert poll. Now, we get a first idea of what exactly causes the adverse effects: too low tariffs coming with low revenues and a lack of flexibility.

Regarding the economic performance of a smart grid, the drivers overweigh the barriers in numbers. The most frequently mentioned driver is the high benefit of electrification in general. Considering the severe ener-

gy poverty in Mozambique and the importance of electrification for economic development, investments in a better energy supply are expected to
yield high benefits. "High benefits of smart solutions" are mentioned five
times as a driver and further specified as potentials to reduce transmission
losses and cost reduction. The cost-benefit ratio of smart grids is assumed
to be highly in favor of the benefits by most of the experts due to the high
perceived relative economic advantage of smart solutions. According to
the representative of Geneva Institute for International and Development
Studies, there are consequently significant potentials for leapfrogging in
favor of smart grid technologies. In contrast to a concept which expects
leapfrogging simply due to the bare existence of a new technology (mentioned above), this statement also takes into consideration the economic
dimensions of leapfrogging: Only if smart energy solutions achieve a better benefit-to-cost-ratio than conventional technologies, it would be economically rational to pursue leapfrogging and implement a smart technology in Mozambique.

Despite the expected high benefits in the future, especially investors
with a strong time preference, focused on benefits in the present, might be
reluctant to implement a smart grid due to higher initial costs. Considering
that a remarkable "short-term orientation" is mentioned as a barrier by two
of the respondents, the higher initial costs of a smart grid if combined with
a high time preference can be a threat to smart grid implementation in
Mozambique. Consequently, the "higher costs" of a smart grid are named
a barrier by four of the respondents. This finding corresponds with the innovation diffusion theory by Rogers (2003), discussed in chapter 0. High
initial costs tend to influence the perceived ratio of expected benefits and
costs of adoption negatively. Since a cost-benefit ratio which is perceived
as unattractive by potential adopters, tends to slow down the rate of adoption, high initial costs can impede an innovation´s adoption (Rogers 2003,
233).

As one possible reason for relatively high costs of smart grid implementation in Mozambique, indivisibilities of grid components in combination
with the low density of population are brought up by some of the experts.
Like with most infrastructures, the size or capacity of grid components
cannot be continuously adapted to the number of consumers: Some components have to be implemented at a certain minimal size, no matter if
there are many or only a few customers. Therefore, in the case of indivisibilities, paired with a low concentration of customers, the costs of infra-

structure are distributed across only a few individuals. This unfavorable relation leads to higher average costs and a lower benefit-cost-ratio.

At first glance, a difference arises between the quantitative and qualitative results regarding the maintenance of infrastructure in Mozambique. While *maintenance* did not produce results that qualified it as a barrier in the quantitative evaluation, a lack of maintenance was brought up as a barrier five times in the interviews. An explanation might be that in the expert poll respondents were more generally asked about the "expected quality of grid management (operation and maintenance) in Mozambique" while in the interviews, respondents narrowed the issue by bringing up maintenance specifically. Furthermore, also in the poll, the variation of results was relatively large for *maintenance,* including a noticeable number of respondents who experience a very low quality of maintenance for Mozambique. It is possible that the qualitative research covered especially experts who expect a lower quality of maintenance. Hereinafter, we shall consider the expected maintenance as a *potential* barrier. One reason, why maintenance might be sometimes incomplete is, how EDM´s staff points out, that workers constantly face a lack of equipment and tools, partly caused by a lack of funds but also due to organizational shortcomings and adverse priorities of EDM´s decision makers.

The goal of this chapter was to outline the main findings. It aimed at constructing a first superficial picture of the issue of interest – drivers and barriers to smart energy in Mozambique. In such a dense chapter, a large part of the variables cannot be discussed in detail. The following chapters aim to address the remaining gaps in analysis. Drivers and barriers will be further decomposed and analyzed in order to reveal the underlying dynamics which form them.

This chapter showed that it is worth to take a closer look at the following factors as potentially important influences to smart energy in Mozambique:

- Ability and willingness to pay,
- Market power, tariffs and revenues, transaction costs and economic environment,
- Technological parameters of smart infrastructure, economies of density, costs and benefits of smart energy as well as grid management,
- Goals and political performance, political and violent conflict, public institutions, regulatory framework, international cooperation, development assistance, acceptance and stakeholders.

A deep analysis of these factors will be presented in the following chapters.

5.3. Finance

5.3.1. Ability to pay

In the preceding chapter, a lack of available capital in general was ranked among the most severe barriers to smart energy implementation in Mozambique. Capital for infrastructure investments can originate from different sources. If the investor is a power utility it can acquire capital from its customers. The more customers are able and willing to pay, the more the utility can increase its prices and enlarge its available funds for investments. Also, investments can be financed by equity capital from prior income from energy markets or completely different markets. Furthermore, capital can be borrowed from financial institutions.

During the qualitative interviews, the lack of capital was an important subject, brought up and explained in length and detail by five of the experts, while both scientists explicitly state that shortness of capital is the central bottleneck of infrastructure investments in Mozambique. It is added by the GIZ representative that the acquisition of borrowed capital is especially difficult for investments involving new technologies in Mozambique. As one reason, he points out the reluctance of banks to put money into new markets in Mozambique, supposedly due to a distinct risk aversion.

EDM´s staff stresses that one of the consequences of capital shortness is that EDM has severe problems financing its ambitious goal to increase the grid coverage to 50% of the population until 2023. Estimates of the World Bank (2015) show that in order to reach this goal, broadly only a quarter or less of the necessary yearly investments are covered by EDM´s own funds, grant aid, donor loans (both ongoing and pipeline) and budgetary support. The remaining financing gap has to be filled if the infrastructure, necessary to achieve 50% grid connection – especially transmission networks and dispatch systems – is to be implemented. Especially this poor financial foundation of EDM´s 50%-goal leads one of the interviewed representatives of EDM to call the utility´s planning "very ambitious". As the respondent states, projects usually turn out to be significantly more expensive than planned.

In its latest energy strategy, even the Mozambican Government acknowledged that the low per-capita income and the resulting shortage of capital for investments are serious obstacles to the improvement of the population´s access to modern energies (Government of Mozambique 2009). The GDP per capita in Mozambique equals to only US$ 529 and according to the Mozambican Ministry of Economy and Finance, about 50% of Mozambicans live below the poverty line, in some provinces even 60% or more (Ministério de Economia e Finanças 2016, 10). Due to the high proportion of extremely poor households, consumer´s ability to pay is expected to be low. A low ability to pay negatively affects energy suppliers´ revenues. Low revenues and profits can hinder energy utilities´ investments.

The existence of support schemes for smart energy implementation could compensate this lack of market revenues (Painuly 2001, Welsch et al. 2013). Support schemes can produce incentives to adopt an innovation by increasing the degree of its relative advantage (Rogers 2003, 236). However, according to several experts, such support schemes practically do not exist in Mozambique. Especially the representatives from the private sector complain about missing incentive schemes to bolster electrification.

Probably, poverty traps contribute to the low ability to pay among large parts of the Mozambican population and the resulting lack of domestic capital. A poverty trap is a mechanism by which societies are kept persistently poor because people below a certain threshold of income or assets have to use all or the larger part of disposable resources for assuring a sufficient level of consumption. Consequently, resources cannot be used for capital accumulation which is necessary for economic prosperity (Sachs 2004). Empirical evidence for Mozambique indicates the existence of poverty traps, especially in rural areas which is argued to be a reason for a persistently low endowment with productive assets and disposable capital (Giesbert, Schindler 2012).

Apart from these economic reasons, poverty traps can also originate from psychological factors. If poverty shapes the socialization of a large share of individuals in a society, this society might develop a value structure which impedes economic and social development. This "culture of poverty" as *Lewis* (1966) puts it, is characterized, for instance, by apathy, short-term orientation, resignation, dependency and a lack of motivation. If a set of persistent and deeply rooted values, shaped by poverty, is cultivated in a society, the individuals can be limited in developing their capacities. What results is a "vicious circle of poverty" (Kersting 1996, 60) which can lead into a persistent lack of capital. Even though the vicious circle of poverty is not necessarily unbreakable (Kersting 1994, 50 and Kersting 1996, 65-68), poverty shall be kept in mind as a potential factor that shapes peoples´ mindsets such that their social and economic development is impeded.

Besides poverty inside the country, Mozambique´s position in global trade is another source of missing capital. In 2016 and 2017, Mozambique´s balance of trade showed broadly US$ 6 billion (UN 2016 c, UN 2018) less exports than imports. This negative balance of trade further aggravates the shortness of capital. Since imports have to be paid for, an excess of imports means that more money leaves the country than payments for exports flow back into the country. A negative balance of trade, therefore, results in net capital exports which means that capital, remaining in the country, decreases.

In the qualitative interviews, experts specified the barrier "lack of capital", pointing out that despite budgetary support from donors, a severely poor endowment of the public sector with financial resources exists. According to the respondents, this lack of capital in the public sector has a negative effect on the electrification of the country. The relevance of a lack of public capital becomes obvious, remembering that experts classify EDM as a typical government-backed monopoly. That is, to a large extent, the company, responsible for extension and improvement of power supply, depends on poor public financial resources.

Furthermore, experts state that power tariffs are too low to accumulate sufficient capital for EDM´s necessary investments. Eventually, a chain of adverse economic conditions emerges: The government sets power tariffs at a too low level which leads to low revenues and a lack of capital in EDM´s accounts. Consequently, the government has to back up EDM but suffers from low funds itself. In the end, EDM lacks capital for new investments. The representative of Geneva Institute of International and Development Studies emphasizes particularly this problem, saying that EDM not only lacks sufficient capital for smart grids but is simply not able to pay for a relevant grid extension at all. The Mozambican government, too, considers the shortage of financial resources in the public sector problematic and calls it one of the barriers to improving the population´s access to modern energy (Government of Mozambique 2009).

A chance to relieve the lack of domestic capital and to compensate capital outflows is seen in the continuously high rates of GDP-growth in

Mozambique. Eduardo Mondlane University´s professor of renewable energy argues in this sense when the respondent states that the constant economic growth in Mozambique will make external funding less necessary. According to recent numbers issued by Mozambique´s National Statistic Institute, real GDP growth can be expected to accelerate again with projected rates of 4.1% for 2018 and 4.7% for 2019 (Caldeira 2018 b).

The ability to pay for smart grids does not only depend on domestic capital. Foreign capital – already identified as a driver – might be able to contribute to closing financing gaps for smart grid investments. Mozambique receives relatively large inflows of foreign development aid, foreign direct investments and remittances (financial inflows from the Mozambican diaspora).

Development aid to Africa amounted to approximately US$ 54 billion net inflows in 2014 and decreased to about 50 billion in 2016 (OECD 2018), of which 2.1 billion official development aid were received by Mozambique in 2014 and 1.53 in 2016 (OECD 2016 a, OECD 2018, 6). Aid numbers decreased in effect of severe concerns regarding the government´s integrity (see chapter 5.6). In the 3-year average from 2014 to 2016, Mozambique became 8th among the largest recipients of official development aid in Africa with 4% of the total aid of all recipients (OECD 2018, 6)

Mozambique has been the number one destination in East Africa for foreign direct investments from 2011 to 2014 with net inflows ranging from US$ 3.5 to 6.1 billion per year (UNCTAD 2016 a). For reference: FDIs per capita and year between 2011 and 2014 have amounted to US$ 588.7 in OECD-countries and to US$ 200 in Mozambique (UNCTAD 2016 a, OECD 2013, CIA 2016).[17] Due to political tensions and economic insecurities, foreign direct investment dropped continuously after 2013 and reached a bottom in 2017 with US$ 1.2 billion. However, pushed by significant investments for the off-shore exploitation of natural gas along Mozambique´s coastline, Mozambique´s National Statistic Institutes expects foreign direct investments to go back up again with expected values of US$ 2.8 in 2018 and 5.7 billion in 2019 (Caldeira 2018 b).

Apart from development aid and foreign direct investments, transactions from the Mozambican diaspora are an important inflow of capital.

---

17   Estimations of population for OECD from 2013 by OECD (2013) and for Mozambique from 2015 by CIA (2016).

Personal remittances to Mozambique moved from US$ 168 million in 2010 to a peak of US$ 233 million in 2013. In the following years, remittances fell to US$ 113 in 2016 and US$ 114 in 2017 (UNCTAD 2018). Despite the recent decrease, remittances can be an important capital source for individual households to invest in energy technology. In contrast to significant parts of foreign direct investments and development aid, remittances are directly addressed to the people in the country of destination.

In line with these numbers, Eduardo Mondlane University´s professor of renewable energy points out that not only in energy, but in nearly all economic sectors in Mozambique, it will depend on inflows of foreign capital if infrastructure projects can be realized since domestic capital is insufficient. For donor money, the majority of the experts´ statements is clearly optimistic. Asked whether donor funds can be expected to be available for investments in Mozambique, the GIZ representative states that donor funding brings about large potentials. According to the respondent, donor-money is "easy money" if projects present elaborate planning. The respondent also underlines that *"We are not talking about peanuts here. There is extremely a lot of [donor]money"*

FUNAE´s respondent´s assessment supports the GIZ representative´s statements. FUNAE´s representative states that donor-money can fill some of the financing gaps, discussed before, since donors usually have financial possibilities, the local partners do not have. Specifically from an aid-recipient's point of view, the representative of the Ministry of Mineral Resources and Energy reports that in his experience, many projects in the energy sector could only be realized due to financial support from donor agencies.

Nevertheless, what remains is the finding that the lack of capital – especially domestic capital – is a major bottleneck to infrastructure projects in Mozambique in general and therefore to smart electrification in particular. The availability of foreign capital can relieve this barrier to some extent, though. Especially donor-funded projects appear to be a promising possibility to introduce smart solutions into the Mozambican power sector. Further research can deepen the understanding and the nature of capital shortness in Mozambique. A detailed analysis can show, where exactly (companies, agencies, government, donors, banks) and in which quantities capital for energy infrastructure investments and smart energy is available or not. For this end, an in-depth analysis of the Mozambican capital market and of the liquidity and capital endowment of potential investors is required.

5.3.2. Willingness to pay

Whether a sufficient willingness to pay for smart grid projects in Mozambique exists, depends on economic objectives and on other alternatives to invest a given budget. If the economic goal is to improve access to electricity, alternatives are quite narrow: only close substitutes of a smart grid infrastructure, e.g. a conventional grid, can successfully contribute to an improved electricity access. On the other hand, when the goal is simply to generate the highest possible return on investment, there is no limitation regarding the area of investment. For both objective functions, it depends on the characteristics of alternatives whether smart grids appear worth to invest in.

Breaking it down to economic concepts: If there is a relevant demand, that is if there is relevant willingness to pay for smart grids in Mozambique, depends on the expected benefits, the *relative price* of a smart grid and on the *cross-price elasticity of demand.*

The relative price of a smart grid is the quotient of the smart grid´s price and some other good´s price. The relative price shows, how many units of the other good must be forgone to acquire one unit of the smart grid (Jehle, Reny 2011, 48). The relative price is a valuable indicator, because rational investors would rather look at how much a smart grid investment would cost in comparison to other investments with comparable outputs instead of looking only at the isolated absolute price.

The cross-price elasticity of demand is defined as a measure for the reaction of the demand for a certain good – here smart grids – to a change in price of another good (Mankiw Taylor 2008, 114). For example, one might be interested in how much the demand for smart grids is expected to increase if the price for alternative investments increases. If cross-price elasticity is very high, investors see promising substitutes for smart grids: Already a small relative price increase of smart grids would ceteris paribus lead to a significant shift to alternative investments. In this case, investors only prefer a smart grid over alternative investments, if the high cross-price elasticity is offset by a relative price very much in advantage of smart grids. It is the purpose of this chapter to take a detailed look at how the experts expect investors to be willing to pay for smart energy in relation to alternative investments.

A low willingness to pay was classified as one of the barriers. As main reasons for this indication, experts mention different priorities, concerns regarding the applicability of smart grids in Mozambique and an insufficient expected return on investment among potential investors. GIZ's representative stresses that reducing energy poverty is not specifically a relevant goal of investors in Mozambique since they did not come with a philanthropic motivation but generally followed commercial objectives. The respondent infers, that commercial investors would only enter the Mozambican smart energy market if this technology had already proven economically viable in this country. According to the representatives of GIZ and the two scientific institutions, economically risky pilot projects cannot be expected to have a promising position on commercial investors' agendas because in Mozambique, there are still so many other areas which promise large benefits, such as the exploitation of the vast natural resources. FUNAE's representative adds that the financial crisis which began in 2007, still influences investors in Mozambique, keeping them away from risky economic adventures. GIZ's representative even goes as far as to state that in the Mozambican electricity sector, private foreign capital other than donor capital is practically not available if new technologies are concerned, especially for off-grid solutions. In this sense, risk aversion can lead to inefficient underinvestment in new technologies (Welsch et al. 2013).

The representative of Geneva Institute of International and Development Studies underlines that for expensive technologies like smart grid components which only amortize in the long run, long term orientation in economic planning is a prerequisite for the existence of a willingness to pay for these technologies. According to the representatives of GIZ and the scientific institutions, time preference is a variable performing in disadvantage of smart energy investments in Mozambique. Behavioral theory assumes that individuals display a certain level of time preference. Hence, for a finite number of periods and a positive interest rate, future utilities are discounted by rational individuals (Samuelson 1937). Strong discounts on future benefits hinder investments with high initial costs, even if they are beneficial in the long run. Therefore, a lack of willingness to pay for smart grids in Mozambique is probably aggravated by the short-term orientation of investors.

Some of the experts, however, challenge or at least qualify the statement, that willingness to pay for smart grids in Mozambique is generally very low. EDM's manager for projects and financing points out that it depends on the way, projects are designed. If, for example, a joint venture approach was applied, using a combination of public capital, aid and private capital, risk could be shared and capital requirements for each shareholder reduced. In his experience, such approaches of burden sharing can be successful in Mozambique, especially when it comes to implementing new technologies.

As far as donors are concerned, experts expect a high willingness to pay for smart energy projects. GIZ´s representative states that donor funding brings about large potentials. Accordingly, donors can easily be attracted as long as projects present elaborate planning. The respondent adds that even if there is "no market", that is, the combination of cost structure and demand is not promising to yield any benefits, donors tend to be willing to support pilot projects. This finding is especially relevant for introducing new technologies into the market since learning curve effects (Fritsch 2014) can be expected to originate from experience acquired in pilot-projects. Eduardo Mondlane University´s professor of renewable energy specifies that pilot projects can increase security for investments, increase efficiency and deliver information about the performance of smart grids under market conditions. By identifying best practice examples, the re-spondent states, reluctance among investors can be reduced, willingness to pay increased and even risk-averse investors attracted to enter the Mozambican smart energy market.

The representative of the Ministry of Mineral Resources and Energy agrees that the willingness to pay among donors for new and smart tech-nologies and their high priority to relieve energy poverty are drivers to smart grid implementation in Mozambique. The respondent tells that some projects, using first smart appliances have already been implemented in a cooperation with donor agencies in Mozambique. For example, the repre-sentatives of FUNAE and the Ministry of Mineral Resources and Energy highlight projects which apply remote metering and mobile payment sys-tems for smart off-grid electricity. As FUNAE´s representative underlines: *"Smart metering – it is not only on-grid. It is also off-grid!"*

In these smart off-grid approaches, customers can purchase a certain amount of energy from an off-grid solar plant using pre-payment via mo-bile banking on their phone. Using information about the time of the transaction, the amount of electricity bought and the amount of energy produced, software can construct consumption patterns which helps the supplier – in this case FUNAE – to optimize and manage the generation capacities. From the fact that these first approaches of implementing smart energy solutions have already been launched, the chief of department de-rives that especially among donors and FUNAE, obviously, some willing-ness to pay exists.

As diffusion theory and corresponding empirical findings show, innovations which can be tried and evaluated in small scale pilot projects generally diffuse more easily. The barriers for a small-scale implementation are smaller than for innovations which directly have to be implemented on a large scale. Furthermore, gained experiences can facilitate re-invention and increase compatibility which facilitates the implementation of the innovation on a larger scale. Therefore, trialability was derived as a core driver for the adoption of an innovation (Rogers 2003, 177). That is, the possibility to test smart grids in pilot projects has the potential to increase their rate of adoption and consequently the adopters´ willingness to pay.

Eduardo Mondlane University´s professor of renewable energy brings up that willingness to pay for smart energy projects can be increased if the potentials and expected benefits of smart energy solutions are communicated more successfully and spread to donors and other external sources of capital. The potentials of a smart energy sector were already extensively described in chapter 0. Among the promising characteristics of smart grids are improved demand side management, more sophisticated control and forecasting, as well as an effective prioritization of loads, e.g. for hospitals or other crucial infrastructure. If these advantages of smart grids are effectively used, the Mozambican energy sector can be made safer, more cost-effective and more environmentally friendly. Technical and distribution losses can be reduced and increased flexibility can help avoid peak-load pricing, outages and load shedding (Welsch et al. 2013). A better harmonization of supply and demand can facilitate the integration of renewable energies. Thus, emissions-intensive energy sources and expensive generator-solutions – often used in Mozambique´s rural areas – can be substituted. A better control of energy flows can furthermore reduce power theft, which is a major problem in Mozambique (Ahlborg, Hammar 2014).

The possibilities coming with a smart grid to increase the scope and quality of energy supply and the contribution to mitigation of climate change go together with typical donor goals. As already mentioned at the very beginning of this work, the United Nations´ Sustainable Development Goals – the leading objectives of the international community to react to current challenges and to meet the needs of the poor – include climate change mitigation (Goal 13) and "access to affordable, reliable, sustainable and modern energy for all" (Goal 7) (UN 2015). The Mozambican government claims that one of its prior goals in energy policy is to improve power supply and the expansion of the grid network (Government of Mozambique 2015). Consequently, willingness to pay might fur-

ther increase – especially among donors and public actors – if consciousness about these advantages of smart solutions for public welfare increase.
The willingness to pay of donors and of the government can also increase
if investments into smart grid infrastructure conjoin with the *creation of
jobs*. Workforce will be needed for construction, operation and maintenance of new energy infrastructure. Additionally, a potentially higher
quality of energy supply due to smart grid investments can improve economic activity and increase domestic labor demand.

All things considered, a lack of willingness to pay is regarded as one of
the impediments to smart electrification in Mozambique. Especially private investors are kept away due to risks to investment, a lack of experience about this new technology in the challenging Mozambican market
and probably also due to time preference. It can be expected that private
investors switch from smart grid projects to more promising alternatives
quite quickly if difficulties occur. Therefore, smart grids can be expected
to only receive funding from commercial investors if their relative costs
and their relative benefits make them economically more feasible than
substitutes. However, the lack of willingness to pay can be relieved to
some extent by donor involvement: experts expect donors to have a relatively high willingness to pay for smart energy solutions. Already existing
and future pilot projects can deliver learning curve effects and increase
willingness to pay among all kinds of investors.

These results shall be seen as a first qualitative tendency regarding the
willingness to pay for smart electrification in Mozambique. Sociology and
economics have come up with several very sophisticated methods for determining and quantifying the willingness to pay more precisely. These
methods can form the base for a deeper analysis of the willingness to pay
for smart grids in Mozambique.[18]

---

18 For further information on willingness to pay measurement see Rao (2009) and
    Endres (2013).

5.4. Electricity market

5.4.1. Market power

Experts from all areas included in the interviews agree that the Mozambican electricity market is characterized by a high level of supply side market power. Government-run EDM is the only relevant player for transport, distribution and commercializing (Ahlborg, Hammar 2014, 118-120). For the operation of the grid, market power is unavoidable since grid infrastructure is necessarily a natural monopoly.

A natural monopoly exists if one producer can supply the whole quantity of the product at lower costs than a combination of several producers, each producing a certain share of the same quantity (Baumol 1977, 810). Under these conditions, competition leads into a situation where only one supplier remains in the market (Meyer et al. 2007). A typical reason for a natural monopoly in grid infrastructure are decreasing average costs due to fixed costs degression and economies of scale (Sharkey 1982, Fritsch 2014).

If a natural monopoly is contestable, that is, if it can easily be replaced by a competitor, the incumbent monopoly is disciplined by *potential competition* and thus, market power is regulated. Barriers to market entry, however, increase with the sunk costs of the monopolistic bottleneck. In the case of very high sunk costs, the incumbent monopoly is considered highly irreversible. Sunk costs are costs which arose in the past and which are irreversible in the short and medium run. Consequently, sunk costs are no opportunity costs of production because they would not vanish, even if production was ultimately stopped (Baumol et al. 1988, 280). The investment into power grid infrastructure is an example for sunk costs: From an incumbent energy company´s point of view, the costs for building its grid infrastructure are irreversible. If power supply was ultimately stopped, the incumbent company could not sell its infrastructure to another company to recover the costs because there is no alternative use for the grid system but power supply.

For new competitors, infrastructure costs are not sunk. To enter the market, new competitors would have to finance the necessary grid infrastructure. However, the potential competitors could also use their funds for alternative investments but grid infrastructure – an option, the incumbent company with its existing grid infrastructure does not have any more. Consequently, new competitors consider costs which the incumbent producer does not need to consider. It becomes evident that a power grid monopoly is not threatened by potential competition. The costs of market entry – building the power infrastructure – are very high for new competitors but sunk for the incumbent monopoly.

Subadditivity of costs in large parts of power transmission and specific investments, combined with high sunk costs, constitute the grid as a highly irreversible natural monopoly. However, the other parts of the value chain in the electricity sector – e.g. generation and commercialization – can be organized competitively and diversified (Fritsch 2014). Furthermore, new competitors could install new grid infrastructure in areas where EDM does not operate a grid system so far and connect formerly unconnected households. According to the representatives of the private renewables company, GIZ and Eduardo Mondlane University, especially decentrally managed mini grids can be a promising infrastructure to put competitive pressure onto the natural monopoly.

In Mozambique, not only transmission but also generation and commercialization of electricity are organized monopolistically to a large extent. Apart from a minor competitive fringe of small suppliers, the hydropower plant of Cahora Bassa on the Zambezi River dominates the Mozambican power generation (see ch. 2.2). As already stated in chapter 2.2, EDM is the major shareholder of "Hidroeléctrica de Cahora Bassa" with 92.5% (World Bank 2015, 19). EDM´s vertical integration leads to an elimination of the price mechanism as a measure of coordination between the value creation stages (Coase 1937). Thus, the market power is extended from the natural monopolistic bottleneck to the other stages of the value chain. Unbundling, that is separating the other stages of the value chain from the monopolistic bottleneck has the potential to enhance competition because potentially competitive stages of the value chain could be taken over by other companies (Gugler et al. 2013).

In EDM´s case, however, unbundling is not realistic, considering the strong protection of its monopoly position by the regulatory agencies. Experts state that EDM is protected by the government holding back competition from the state-owned monopoly. However, some experts add that some progress has been made regarding the openness of the electricity market to new participants. According to EDM´s manager for projects and financing, until a few years ago, the protection of EDM by the government was much stronger than today. Nowadays, new laws supported market entry of private companies. Nevertheless, shifting the focus from the formal legal situation to actual performance of government and EDM, no real willingness to increase private sector engagement exists among both government and EDM, as representatives of the scientific institutions, GIZ and even one of EDM´s representatives claim.

The representative of Geneva Institute of International and Development Studies points out, that in electricity markets, there are limits to liberalization. Central decision-making avoids duplication and incoherence and can therefore accelerate grid extension. Thus, a central goal in emerging economies – to quickly enhance access to electricity – could be achieved more effectively if there was one utility with a clear responsibility and the power to enforce its decisions. Hence, it depends on the performance of the actual public energy utility and the corresponding regulatory agencies if improvements for the population are achieved or if the market-dominating position is only misused to keep away potential competitors.

Summarizing the experts´ assessments, EDM strongly concentrates the market power across all parts of the value chain. It is a practically irreversible monopoly and benefits from a back-up by the Mozambican government. The small competitive fringe of independent generation companies and off-grid solutions does not have significant regulating effects on EDM´s performance.

Which consequences does this strong concentration of market power in the Mozambican energy sector have? As chapter 5.1.3 has already shown, a lack of competitiveness usually leads to allocative and dynamic inefficiencies. Some of the consequences are too high prices, too small market volume, an inefficient level of investment and too little development of new technologies (Jehle, Reny 2010).

However, usually there are possibilities to restructure a market and thus, to tackle market power. Potential or actual competitors can reduce market power and contribute to a decrease of the price and an increase of quantity. Small producers in a competitive fringe have to accept the monopolistically set price and adapt their produced quantity according to the given price. Thus, the total quantity of the product in the market increases and in the case of a typical demand function, the marginal willingness to pay of the consumers decreases. A lower marginal willingness to pay means that consumers would not accept the previous price any more and the price level decreases. In effect, demand is partly supplied by the competitive fringe whereas the larger share is still served by the incumbent monopoly. Since market volume has increased and the price has fallen, the market outcome in a market with upcoming competition comes with lower welfare losses than in a completely monopolistic market. The larger the market share of the competitively structured fringe becomes, the closer the allocation gets to the welfare-maximizing solution (Meyer et al. 2007).[19] In this process, the volume of sold power would increase. Eventually, this increase of power sales would make new grid infrastructure necessary as more and more households would be connected and grid capacity would have to increase. In this process, new grid infrastructure could be a smart one.

Competition also addresses the quality of the product. In order to attract customers, competitors are constantly part of a race for better quality. The companies typically regard a distinct quality of their product as a means to increase their share in the market. Talking about energy, this quality-enhancing competition can improve energy security, reduce blackouts and load shedding and speed up the introduction of new technologies like smart grids. Different electricity products like flexible tariffs, a broader choice of contracts or differentiated power products, such as electricity, exclusively from renewable sources, are further possibilities for a quality-driven differentiation of the market. A lack of competition, though, can lead to a deterioration of the quality of supply.

Lacking competition not only has consequences for market incentives, allocative efficiency and corporate culture but can also harm the resilience of power supply. The low market share of companies other than EDM leads to a low diversification of generation and a very centralized decision making in the electricity market. According to EDM´s manager for grid and transport, this situation brings about stability problems since failures or blackouts in EDM´s infrastructure cannot be compensated by other suppliers.

The concentration of market power in the Mozambican energy market does not remain without regulatory intervention. In order to reduce the exploitation of market power, the Mozambican government performs a price cap regulation on power prices (World Bank 2015, 18). A price cap aims to prevent exploitative monopolistic price setting. The regulation forces the monopoly to accept an upper boundary for the price, lower than the monopoly price which produces the profit-maximizing equalization of marginal costs and marginal revenue (Cournot 1924). In effect, the price-cap narrows the monopoly´s profit, brings the allocation closer to the effi-

---

19   For formal reasoning see Meyer et al. (2007).

cient scenario and reduces the welfare losses. It has to be considered, though, that as long as the price cap does not exactly equal the efficient price, the monopoly exploits its possibilities for profit maximization and welfare losses result. However, a continuing decrease of the price cap in order to abolish all exploitation of market power comes with a serious risk that the price-cap falls below the average costs of the supplier. In this case, the supplier generates losses and will eventually be forced to leave the market unless it receives subsidies to cover the deficits.

The price-cap regulation might tackle one problem of a monopolistic market – too high prices – but it can at the same time make the presence of market power more permanent. The reduced price can keep away new companies, perhaps more innovative and more interested in new technologies than the incumbent monopoly. Thus, the price-cap regulation might effectively tackle market power in the short run but make the market power even more permanent in the long run.

Among the experts, especially the entrepreneur as a representative of the private sector criticizes EDM´s domination of the Mozambican electricity market. The respondent underlines that the missing logic of competition causes significant incentive problems for economic actors: EDM had incentives to exploit its market power for profit-maximizing and potential investors were reluctant to enter the Mozambican electricity market due to the dominating position of EDM. GIZ´s representative specifies that EDM often denied access of potential competitors to essential facilities like grid infrastructure, thus reflecting the typical behavior of an incumbent monopoly, willing to drive competitors out of the market. Even the three employees of EDM acknowledge that the lack of competition is a problem for the Mozambican energy market. They all agree that more competition would put more pressure on EDM to improve its services and its performance.

Besides adverse economic incentives, arising from EDM´s position as a monopoly, experts also see its corporate culture as one of the barriers to a better power supply. An organization´s culture is the normative system of shared values and beliefs, shaping how the organization´s employees feel, think and behave (Schein 1990). The culture of an organization constitutes strong path dependencies as the organization operates according to its cultural identity. A company´s cultural identity originates from the conventional behavior, norms, values and mental maps established and preserved in the organization over time. It forms the company´s "memory" (O´Keefe 2002, 133).

EDM is clearly described as a conservative organization by the representatives of the Ministry of Mineral Resources and Energy, GIZ and the private company. Unchallenged by potential competitors, EDM´s reluctance to innovations is not sanctioned by the market and consequently, the pressure to act more innovation-oriented is low.

GIZ´s representative specifies his judgement about EDM´s conservative culture, pointing out that EDM is very reluctant regarding the implementation of new technologies and hesitating when it comes to alternative ways to supply energy, like off-grid solutions or renewable energies. Furthermore, according to the entrepreneur, EDM´s way of thinking is not economic but rather political. As an example, the respondent mentions that instead of building new grids where they achieve the largest economic benefit, EDM quite often follows governmental priorities such as connecting areas with a strong pro-government voter base. Concluding, the experts´ judgements indicate that EDM´s corporate culture is neither innovation- nor business-oriented. The representative of the Ministry of Mineral Resources and Energy states that only more competition could drive EDM to be more open to innovations.

Organizations are typically relatively innovative, if they are learning organizations. The core element of a learning organization is a culture of learning. A learning culture is a culture that values knowledge and innovation and creates a favorable environment for exploration and experimentation (Hamel, Prahalad 1991). In a learning organization, change and the regular implementation of new innovations is self-evident. Therefore, a learning, open-minded organization can implement new innovations much faster, more effectively and with a higher benefit than organizations with a rather conservative culture. Conservative organizations fail to learn due to closed mindsets. Change and the implementation of innovations is impeded by cultural rigidities (O´Keefe 2002, 137).

EDM fulfills some typical attributes which lead to a conservative organizational culture. Especially relatively large organizations with many employees tend to be slow and ponderous, particularly when it comes to technological change. Furthermore, in the energy sector, a constant flow of power for supply security, has a top-priority. What follows is a risk-averse decision making reducing the company´s flexibility: EDM might emphasize routine and security over change and optimization in order to reduce risks that supply security could become even worse than it already is.

An organization´s decision-making structure significantly influences the culture of its operations. For example, centralized decision-making is usually an impediment to innovativeness. EDM´s staff stresses in the interviews that decision-making in their company is organized quite hierarchically. A hierarchical structure might come with stability and continuity but it may resist the implementation and re-invention of innovations (Rogers 2003, 179 and 189). Many minds are usually more innovative than one. Genuine reliance on employee initiative and delegation of responsibilities bolsters innovativeness (O´Keffee 2002, 137). One of the consequences of EDM´s resistance to innovation and alternative electrification is FUNAE´s active involvement in rural electrification. FUNAE, originally founded as a fund, only to finance but not to implement or manage electrification projects, was driven into the off-grid market because EDM was not willing to involve itself in this area – despite the vast lack of rural electrification in Mozambique.

In the last years, some noticeable progress regarding the promotion of competition in the Mozambican electricity market has been made for small-scale investments. The Mozambican government´s latest energy strategy sets the goal, to encourage the participation of the private sector in electricity projects. Especially for generation, private companies are aimed to be attracted to further increase competition on the non-monopolistic stages of the value chain (Government of Mozambique 2009). Off-grid electrification is significantly driven by private companies and development agencies. However, GIZ´s representative states, off-grid electrification in Mozambique often comes with poor cost coverage due to the low ability to pay of poor rural customers. Usually, donor money is needed since other measures of guaranteed revenues, like support schemes from the government, do not exist.

According to the representative of the Ministry of Mineral Resources and Energy, a problem with private sector engagement in the Mozambican electricity sector is that many private companies are curious but have no experience in Mozambique nor in the electricity sector at all. Therefore, many investors were unreliable and quickly drew back from their engagement if difficulties occurred. Consequently, governmental agencies had problems choosing a reliable private partner when they are willing to open tenders for electricity projects.

Taking all things into consideration, despite first steps to increase competition in the Mozambican electricity market, EDM remains the dominating utility. It is highly vertically integrated and extends market power from the monopolistic bottleneck – the grid – to the other parts of the value chain. A price-cap regulation aims to keep prices below the profit-maximizing monopoly price. At the same time, this upper boundary for the price probably also holds back new competitors and consolidates the market power. Furthermore, despite regulatory interventions, the quantity of power sold in Mozambique is still very low. Typical problems of monopolistic markets result, like a poor grid coverage and quality shortfalls, coming as blackouts, bad service and losses. EDM is criticized to hold back private competitors and its conservative corporate culture is expected to make it reluctant to new technologies and innovative electrification strategies. EDM´s reluctance to new technologies and its strong domination of the Mozambican electricity market at the same time are expected to constitute severe barriers to the implementation of smart grids. A reduction of market power either by effective regulatory actions or by new competitors could unlock innovativeness, increase the market volume and make new grid infrastructure necessary – a chance for smart energy.

5.4.2. Tariffs and revenues

Experts regard the bad revenue situation in the Mozambican electricity market as an important barrier to the implementation of smart grids. They state that the revenues which can be achieved in the Mozambican electricity market are not high enough to generate a sufficient surplus given the current cost structure for generation, transmission and commercializing. The strict price-cap regulation of electricity tariffs is seen as the main reason for the bad revenue situation. According to the experts, electricity tariffs are and have been too low in the past years to refinance maintenance and the necessary investments.

Until summer of 2017, power prices in Mozambique ranged from USc 1.4 per kWh for the social tariff to USc 7.55 for the general tariff in the cheapest category (low total consumption) and reached up to USc 11.77 per kWh for the general tariff in the most expensive category (consumption of more than 500 kWh per month and household). For special tensions, domestic-only-use and agriculture, there were special tariffs, not exceeding USc 8.00 per kWh (EDM 2016).

Since the latest price increase in the summer of 2017, the price range starts at the unchanged social tariff but reaches USc 16.85 in the most expensive category. The agricultural tariff was not increased. The domestic-only tariffs changed from USc 5.28 to USc 7.14 in the cheapest category and from USc 7.85 to USc 10.61 in the most expensive category (EDM 2017, Caldeira 2017).[20] As the further reasoning will show, even the increased price level in the Mozambican power market complicates the revenue situation significantly.

As the main player in the Mozambican power market, especially EDM is hit by the strict price regulation. Considering the ambitious targets for grid extension, EDM´s lack of funds constitutes large financing gaps for the necessary investments. The entrepreneur gets to the heart of it, when he says: *"EDM is always just before ruin"*. In a press conference in May 2018, EDM´s chief financial officer, Noel Govene, stated:

> "Today we are at a medium tariff of eight cents per kWh with costs of ten cents. Consequently, mathematics shows us that eight cannot cover 10." (Caldeira 2018 a)

At the first glance, these assessments might seem surprising, considering that it was argued that the price cap is needed to tackle the problem of too high prices as an effect of market power. At this point, a trade-off between allocative and dynamic efficiency becomes obvious. In order to reduce market power and to optimize the allocation, a price-cap regulation is a promising tool. However, the lower price reduces the utility´s profit and therefore limits available funds for investments in grid extension and new technologies. This combination of insufficient revenues despite high market power is an important shortcoming of the Mozambican energy sector.

Eduardo Mondlane University´s professor of renewable energy explains that besides reducing market power, the government justifies its intervention into price setting with social justice. To guarantee access to energy for a large part of the population, low energy costs are desirable. However, it is questionable if the price-capping, coming with a social-political motivation really achieves its goals since only people who are connected to electricity supply benefit from the low electricity costs. The other broadly 70% of the population, who do not have access to electricity and mainly live in very poor rural areas, do not benefit at all. Therefore, the low tariffs might actually counteract effective poverty relief by impeding the generation of funds for grid enlargement and improvement of power supply for the poor.

---

20 All prices computed at a Dollar-Metical exchange rate of 76.42.

Furthermore, increasing the average tariff does not necessarily imply an increase of tariffs for the extremely poor. This view is clearly expressed by FUNAE´s representative. In his opinion, the problem is not the social tariff but the tariff structure in general: Especially wealthy households with a high ability to pay should pay more and shoulder the investment costs for new and smart grids, he says. Another possibility to increase the average tariff is by implementing special higher tariffs for industries with an extraordinary power demand (Government of Mozambique 2009). However, this strategy could interfere with Mozambique´s industrialization-goals.

Already in its energy strategy of 2009, the Mozambican government acknowledged that tariffs should reflect the "real costs" (Government of Mozambique 2009). Despite this goal, a tariff increase, including EDM´s costs, the relatively high inflation (rates between 2.6 and 12.7 since 2010 (World Bank 2016 a)) and a mark-up for future investments could not be enforced until now. The World Bank (2015) calculates that if the government´s goal of electricity grid access for 50% of the population until 2023 is to be met, a tariff increase of approximately 45% (nominally) would have been required in 2015. In reality, though, tariffs were only increased by 26.4% nominally in 2015. During the years before (2010 to 2014), tariffs were not increased which in real terms equals to a deterioration of the price of 20% in that period (World Bank 2015, 43).

However, as mentioned before, in the summer of 2017, EDM increased the power tariffs again, with increases between broadly 26% for domestic use and up to 33.40% for large consumers. The social and the agricultural tariff were not increased. The increase was only partly justified with new infrastructure investment. A large part of the increase was due to increased buying prices for power from the Cahora-Bassa hydropower plant which EDM has to re-buy from South Africa and from private generation capacities (Caldeira 2017). Furthermore, a large part of the nominal increase is set off by the inflation of approximately 14% in 2017 (Folha de Maputo 2016). Therefore, it is highly questionable if the latest tariff increases are sufficient to compensate for the structural losses of the last decades. Nevertheless, already the current increases lead to outrage in some of the media and population (ibid.). Generally, for the Mozambican government, tariff increases for electricity appear to be politically risky. In Mozambique, there is a high probability of social unrest if prices for basic commodities increase. In the last decades, Mozambique regularly faced resistance from civil society and even violent upheavals in the streets of the major cities when the government announced increases of the prices for basic commodities like bread, electricity or water (Behnisch 2010). Thus, increases are limited if no social unrest shall be provoked. Consequently, a sustainable improvement of EDM's economic situation has to look not only at the revenue but also on the cost side. Increased efficiency makes higher tariffs unnecessary. This is where smart grids can contribute. If their potentials are exploited thoroughly, EDM can alleviate its economic pressure by reducing costs and avoid further pressure to increase tariffs.

Besides the adverse tariff structure and the costly, inefficient operations of EDM, further reasons for the bad revenue situation in the electricity market, mentioned by the experts, are technical losses due to defective infrastructures and problematic political measures. For example, EDM is forced by the Mozambican government to supply power for public lighting without direct monetary compensation (World Bank 2015, 28). EDM´s staff also criticizes that the result of EDM´s tendency to follow rather political than economic objectives is that investments were often directed to areas that are politically feasible but not necessarily economically beneficial. Furthermore, EDM states that the increasing introduction of new relatively expensive generation sources, especially renewables, raises EDM´s costs (Caldeira 2018 a).

Experts do not only criticize the low level of the average power tariff but also the lack of variation in the tariff structure. Representatives of EDM, FUNAE and GIZ state that due to a lack of regional price differentiation, tariffs do not reflect the real costs. Electrification is usually economically less beneficial in poor rural areas with a low density of population than in larger, densely populated cities (see also chapter 5.5.2). However, electricity tariffs are the same for all regions in Mozambique. Thus, possibilities for a more inclusive exploitation of the different willingness to pay of costumers in different regions are restrained and potentials for increased revenues closed. Again, there is a dilemma between economic and short-term social objectives: From, an economic perspective, especially in poor rural areas, where the density of population and the endowment with disposable capital are low, high tariffs would be required. Higher tariffs in poorer areas counteract poverty alleviation, though.

Furthermore, EDM does not offer any scarcity adjusted tariffs (EDM 2018). As chapter 5.1.3 has shown, flexible tariffs that charge more when power supply runs short, create incentives to use less power in times of high scarcity by driving consumers with a low willingness to pay out of the market (Schreiber et al. 2015).

The benefits of a smart grid can only be fully exploited if flexible tariffs are introduced. A smart power system collects and processes large amounts of data about consumption, production and load patterns. If this information is used to adjust power prices to the scarcity in different periods, load management becomes more effective. Among the benefits of a better load management are less black-outs and cost reductions. Impeding these potential benefits of a smart grid, the non-existence of flexible tariffs constitutes a barrier to the introduction of a smart grid in Mozambique.

Summarizing the analysis of power tariffs, investments into new and smart grid technologies are impeded by a bad revenue situation in the Mozambican energy market, mainly caused by governmentally controlled too low power tariffs. Furthermore, scarcity-adjusted flexible tariffs do not exist although they can stabilize electricity supply. To exploit the advantages of flexible tariffs, smart grid technologies would be required.

5.4.3. Transaction costs

For Mozambique, smart energy solutions are still a quite new technology. This brings about enhanced needs for information and testing. On the other hand, smart energy is expected to significantly decrease transaction costs by using ICT like more sophisticated software solutions for data processing. Such software solutions improve the availability and evaluation of large amounts of information (EPRI 2011). It was one of the objectives of the empirical analysis to determine which of these two influences on transaction costs overweighs for the Mozambican context.

Experts state that smart energy implementation in Mozambique can be expected to come with relatively high transaction costs: Due to missing transparency among governmental entities, a low quantity and poor quality of statistical economic data and little use of information and communication technology, the availability of necessary information for investors is deficient. High costs for information-gathering result. From a private sector perspective, the entrepreneur adds, responsibilities of the authorities are often unclear. According to the respondent, this lack of transparency and clarity leads to additional challenges for information-gathering.

EDM´s quality of grid management can be expected to be severely harmed by high transaction costs. According to EDM´s manager for grid management and maintenance, EDM´s very centralized administration protracts the exchange of information from and to remote areas. Consequently, the necessary exchange of information inside the company can only be fulfilled to a limited degree, the respondent states. The respondent underlines that processes could run much more efficiently if some decision-making and administration was handed over to the local entities. Transaction costs are further increased since significant risks and challenges of the Mozambican market require additional information to reach a certain level of reliability for assessments of costs and benefits, the representatives of FUNAE and the private company say.

So far, EDM does not collect any sophisticated data about demand patterns of consumers. According to EDM´s staff, information about the clients´ demand patterns would help a lot to optimize supply. The integrated use of information and communication technology could improve EDM´s information-gathering and at the same time reduce its costs. As the representative of the Ministry of Mineral Resources and Energy points out, already existing smart energy initiatives and pilot projects could facilitate the use of the digitalization´s potentials for the Mozambican electricity sector.

Summarized, it can be expected that the introduction of smart energy solutions in Mozambique will cause high transaction costs in the beginning. However, if used effectively, smart energy can contribute to the reduction of transaction costs in the long term by improving information gathering and processing.

5.4.4. Digitalization

A necessary condition for a successful application of smart energy technologies is a sufficient coverage of information and communication technology in line with favorable framework conditions for digitalization. Even though it can be argued that a high willingness and ability to pay for smart grids would also induce investments into the necessary information and communication technology, an already existing ICT-infrastructure facilitates the implementation of smart grids.

It should not be neglected that there are different forms of access to information and communication technology, not only access to the internet. Especially in the Mozambican context – and generally in African countries – much information is gathered and transmitted through the conventional mobile phone network, examples are different forms of mobile money transfer, news, private or professional communication, commodity prices or weather information. However, internet use can be an indicator for the diffusion of digital technologies in a country. Reliable and current data for ICT use in Mozambique is very rare. Consequently, existing data should be treated prudently. Nevertheless, the following numbers for the share of individuals using the internet in the entire population can be one indicator for the state of digitalization in Mozambique. According to numbers from the International Telecommunication Union, Mozambique has increased its share of internet users nearly by the factor six from about 4% in 2010 to about 21% 2017 (ITU 2019). With this development of internet use, Mozambique is highly representative for the development on the whole African continent which shows a growth of the share of internet users from 6.5% in 2010 to 22% in 2017 (ibid.). In the same period, the worldwide share of internet users grew much slower – though on a higher level – from about 30% in 2010 to 48.6% in 2017 (ibid.). A possible interpretation is that Mozambique and other African countries are catching up regarding individual internet use while the digital divide remains significant. In the light of the high rates of adoption of digital technologies worldwide, numbers of individual internet use in Africa and Mozambique specifically have most probably continued rising significantly since the latest ITU data from 2017.

Like in many African countries, the implementation of ICT-solutions in Mozambique is forcefully driven by the quickly expanding cell phone market (INE 2014, World Bank 2014). The boom of the mobile phone market, coming with several mobile applications such as mobile payment by the widespread African mobile banking system M-Pesa has brought large improvements in ICT availability. Nowadays, most parts of Mozambique are covered with mobile internet and there is competition between three large mobile communication companies (Vodacom, MCel, Movitel)[21]. Already in 2014, statistically nearly every Mozambican owned a cell phone (INE 2014), many of which are nowadays smart phones. At the same time, fixed-line internet remains at a low level (World Bank 2014) while smart phone use can be expected to keep on expanding, as several respondents in the interviews state. Consequently, digital solutions

---

21   The mobile phone-companies display information about the coverage with mobile internet. On http://www.vm.co.mz/en/Individual/Products-and-Services/ Coverage, Vodacom offers a map of the country with all covered regions.

for the energy sector or other areas of application will have to be customized for mobile use if relevant levels of application should be achieved.

It became evident in the qualitative interviews that the good coverage of mobile phone network and mobile internet is seen as an important driver for smart energy diffusion in Mozambique by several respondents. Experts attest that Mozambique can score with a good and quickly increasing abundance of mobile information and communication infrastructure. Respondents furthermore mention that the current diffusion of smart energy solutions in Mozambique, on-grid but especially off-grid, shows that digitalization in the energy sector is proceeding. As the representative of the Ministry of Mineral Resources and Energy highlights, the quickly proceeding digitalization of the Mozambican economy and society has the potential to further accelerate the introduction of smart solutions. According to the representatives of EDM and FUNAE, both actors are very open to further implementing digital solutions to improve their services. EDM´s staff clearly demands the digitalization of the company´s processes in order to increase efficiency by lowering transaction costs. The respondents explain that the management of the grid is executed completely manually and without real-time information about the state of load and the quality of infrastructure in different parts of the grid. Potentials for automatization and better flow of information coming with information and communication technologies could significantly improve the quality and efficiency of grid management and operations. For example, the respondents demand a software for EDM that visualizes load factors, fluctuations and damages in the grid. Thus, the lack of information about the performance and functionality of isolated components would be overcome. As one of EDM´s representatives summarizes: *"Digitalization is definitely welcome."*

Not only in Mozambique but all across the global south, digitalization is proceeding. In African societies which are characterized by a strict mobile-first technology use, applications for smart phones are developed rapidly and used vastly. Areas of application range from mobile payment or remote monitoring of smart off-grid systems (see chapter 8.4) to new possibilities of political participation. For example, digitally capable citizens can nowadays use smart phone applications to report technological failures of infrastructure, pay their energy bill or report dubious incidents in electoral processes. Furthermore, social media networks or other forms of online participation increasingly substitute or supplement offline forms of communication and political participation, such that new forms of participation are implemented by civil society ("invented space" (Kersting

2013)). The combination of online- and offline participation constitutes a blended democracy which describes the interaction between online and offline participation and offline and online democracy. Examples for the implementation of electronic participation and electoral monitoring from Uganda and Tanzania can be found in *Kersting, Shayo* (2016), *Kersting, Shayo* (2017) and *Matsiko, Kersting* (2018). Such examples illustrate how digitalization concerns not only technological infrastructure but also the societies and political systems of the global south. Likewise, the assessments expressed by the different respondents in the interviews indicate a high level of openness to technology and digital solutions among the Mozambican population.

Nevertheless, the mainly optimistic statements of the experts regarding digitalization in Mozambique should not take away attention from the fact that there are still obstacles for smart energy diffusion in Mozambique due to missing digital infrastructure. Respondents from EDM and GIZ mention that there are still regional deficits in ICT-coverage.

5.4.5. Economic environment

Economic developments and framework conditions shape the electricity market and influence the potential of new technologies like smart grids. Today, important industrial branches in Mozambique are basic nutritional products, liquors and drinks, chemistry (fertilizers, colors and soap), cement, petrol products and tobacco (Government of Mozambique 2009, 7). Recently, Mozambique also experiences a boom of extractive industries due to the large reserves of fossil resources, such as natural gas and coal (see ch. 2.2). Strong growth rates in the primary and secondary sector are key drivers for services like consultancy, research and development, controlling and finance, to mention only a few examples. A more specialized economy with an increased level of division of labor brings about new markets for information and communication technology.

As mentioned before, Mozambique is characterized by a strong and relatively constant economic growth. The interviewed actors regard this strong growth of the Mozambican economy as an important driver to the electrification of the country and to the implementation of smart technologies. Economic growth in Mozambique – though coming from a low level – brings about progresses in industrialization and increasing productivity in the agricultural sector, both serving as drivers to electrification, respondents say. More and more businesses and farms grow to a size and a level of productivity, even in rural areas, that requires the use of information and communication technology, electric lighting, storage and cooling which all need stable electricity supply.

The growing tourism sector in Mozambique is another source of economic growth and a driver to electrification, even at very remote spots. Despite minor returns of old conflicts from time to time, Mozambique is more and more perceived as a relatively safe country since civil war has been ended already more than two decades ago. Tourists are attracted by the long beaches, rich culture, game reserves and diverse nature (Jones 2010). Another driver, especially for the extension of the Mozambican power grid is the connection of public buildings. EDM, in line with the Mozambican government, is aiming to quickly connect all administration posts in provinces and districts to the grid, EDM´s manager for grid management and maintenance says.

Looking at the labor market, representatives of EDM and Eduardo Mondlane University indeed agree that by *formal* qualification, sufficient staff – like engineers, technicians, economists and administration professionals – is available for energy-related work. Furthermore, EDM´s representatives mention that EDM seems to be an attractive employer since qualified applications constantly exceed vacancies. However, despite the applicant´s qualification "on paper", the representatives of EDM, the Ministry of Mineral Resources and Energy and GIZ emphasize that specialized knowledge about the electricity sector is often missing, especially in the area of new technologies. Additional training would be needed but is often not available. Therefore, the representative of the Ministry of Mineral Resources and Energy specifies that often it is not a lack of wanting but a lack of options for specialized qualification. The respondent criticizes that Mozambican universities do not offer differentiated courses for the energy sector and that EDM, FUNAE and public authorities do not offer sufficient options for targeted continuing education. According to the respondent´s experience, there is quite a high level of frustration among the ministry´s and the utilities´ staff due to a lack of possibilities for additional training, both in academic and vocational education. In line with EDM´s staff, the representative of the Ministry of Energy and Mineral Resources argues that the lack of specialized knowledge and specialized training is also a result of missing funds.

The lack of specialists is even aggravated, as the chief of department experiences, by a critical brain drain: A part of the very well qualified staff leaves the country or stays abroad after finishing studies in foreign countries. One of the main reasons, the respondent says, are the relatively low salaries compared to international standards in the Mozambican electricity sector. As the analysis of the determinants of innovation diffusion in chapter 0 showed, a lack of knowledge can impede innovation diffusion severely. The experts´ statements imply that especially principles knowledge is missing among Mozambican energy sector employees, that is, knowledge of the functioning principles of how a technology – here energy infrastructure and smart grids – works (Rogers 2003, 173). Principles knowledge is basically an issue of basic education in schools and universities. A lack of principles knowledge is therefore an indicator for an insufficient quality of the Mozambican education institutions. If potential adopters lack knowledge about an innovation but display motivation to handle the new technology, re-invention is likely to occur (Eveland et al. 1977, Larsen, Agarwala-Rogers 1977, Kelly et al. 2000). The innovation is modified such that it fits the adopter´s capacities and preferences. Re-invention brings about the potential to make the innovation more suitable for the local circumstances and to reduce its complexity. On the other hand, re-invention always comes with the risk to disable a technology or to handle it in a suboptimal way.

It has to be considered that in a country like Mozambique with a persistently poor society, one could expect a "culture of poverty" (Lewis 1966, see also chapter 0) which makes people fatalistic, short-term oriented and uninspired, such that innovative behavior like re-invention is not to be expected. FUNAE´s representative expresses that the Mozambican economy generally misses initiative of economic actors. According to the respondent, innovative power of economic actors in Mozambique is low and a generally hesitant economic culture keeps people back from economic challenges or risks, necessary to realize promising visions for business. However, empirical findings by *Kersting* (1996) indicate that there is no deterministic correlation between persistent poverty and missing innovativeness. In contrast, the author finds that in marginal settlements, inhabitants are often quite innovative and willing to experiment with new technologies (Kersting 1996, 65-68). It should be considered, though, that re-invention – especially if based on a lack of knowledge – also comes with the risk to make the technology less effective as not all its potentials are exploited and not all functions used.

Taking the focus away from human qualification, another severe barrier to economic development in Mozambique is the poor state of transport infrastructure. Only a few roads connect the different parts of the country and if existent, roads bother drivers with frequent potholes or come as dirt roads without any asphalt pavement. Railways are few in numbers and only allow for relatively slow velocities. The representative of the Ministry of Mineral Resources and Energy states that grid maintenance is severely harmed by the deficient infrastructure since transport costs are high if components have to be repaired or new lines are to be constructed.

Changing the topic again from the barriers to the drivers, Mozambique has large potentials for renewable energies such as hydro, solar and wind power, as chapter 2.2 already showed. Scientific experts as well as representatives from the private economic sector mention that especially for off-grid and mini-grid solutions, renewable energies, especially solar and hydro, are vastly used and can be a driver to electrification. EDM´s representatives state that EDM has been reluctant on wind and solar in the past years but aims to increase their share in its generation. However, for grid-connected renewables, the missing capacity and flexibility of the grid can be a bottleneck. As the entrepreneur puts it, the current situation of the grid is "one of the biggest problems". The respondent adds, that a smart grid that facilitates the management of intermittent renewable energies in the power system can create new possibilities for grid-connected renewables. The respondent continues that a stable energy supply is highly necessary because in Mozambique, especially small and medium companies without a privileged grid connection suffer from missing security of supply. Common, though expensive options to provide a stable power supply for one´s company are batteries and generators that can be used in the case of blackouts. More cost-effective solutions are needed to improve energy security, EDM´s manager says and calls smart grid solutions "necessary" to integrate alternative energy sources. The respondent qualifies that tailored solutions will be needed for urban areas on the one hand and rural zones on the other hand due to big differences in the economic conditions.

In this chapter, it became evident that the economic environment of smart energy implementation in Mozambique contains several drivers but also some barriers. Positive developments are the high economic activity, the new possibilities to use renewable energies and the increasing perception of smart energy as a tool to improve power supply. On the other hand, spatial conditions and transport infrastructure are examples for impediments in the economic environment.

5.5. Infrastructure

5.5.1. Technological parameters

The present Mozambican grid infrastructure was designed for a strongly centralized generation, focused on the Cahora-Bassa dam. The increasing deployment of decentral, fluctuating power generation and increasing energy demands of a growing and digitalized economy are challenges to the Mozambican energy system. Whether a smart grid is a viable option to address these challenges, clearly depends on how well it is technologically applicable in Mozambique. Is the integration of smart appliances compatible with the existing infrastructure? Are necessary prerequisites for the implementation of a smart grid fulfilled? Is leapfrogging a viable option?

What certainly argues in favor of a smart grid in Mozambique, is that it addresses central problems of the Mozambican power sector like significant transmission losses, increasing fluctuations due to the introduction of solar and wind power and a lack of storage capacities. Especially EDM´s staff member, responsible for projects and financing, evaluates the potentials of smart grids very optimistically due to the technological and economic problems to store energy: *"It [smart grid] is the future. Because you cannot really store electricity sufficiently. This [smart energy sector] will be the perfect situation."*

A potential lack of interoperability of smart and conventional grid components is a threat to the introduction of smart grids. Furthermore, especially the distribution of power to the households on a regional or local level has to be adopted to traditional building-techniques and resilient to power theft. It is questionable if advanced and valuable smart grid technologies meet these prerequisites. On the other hand, GIZ´s representative argues, that a large part of the Mozambican grid and its components (transformers, dispatch stations) are relatively new and modern for an African context since large-scale electrification started relatively late. FUNAE´s representative argues in the same sense when he qualifies that smart components are generally applicable in the Mozambican electricity sector, but to a large extent, tailored solutions will be required.

Instead of upgrading an existing grid infrastructure to a smart one, leapfrogging – brought up by the representatives of FUNAE and Geneva Institute of International and Development Studies – can be another option to implement a smart energy infrastructure. Where power grids do not exist at all, leapfrogging could promote smart grids in relation to conventional

131

technologies. In this case, difficulties due to problems with the interoperability of smart and conventional grid components do not occur.

In the case of a smart grid, applicability is enhanced due to the vast possibilities for re-invention. Through re-invention, the innovation can be adapted more appropriately to local and changing conditions (Rogers 2003, 185). What is more, re-invention improves the sustainability of an innovation: Adopters who modify and re-invent a new technology begin to regard the technology to some extent as their own. Thus, the probability that the adopter continues to use the innovation increases. Therefore, re-invention can contribute to building a sense of ownership among the adopters of an innovation (Rogers 2003, 429). Re-invention is encouraged if a technology can be modified in several ways (Rogers 2003, 181). For innovations that are quite complex and difficult to understand, there is an increased possibility of re-invention. In this case, re-invention can lead to a simplification of the technology (Larsen, Agarwala-Rogers 1977).

Smart grids and smart off-grid solutions come with the advantage that they can be applied in many different ways and that they can be adopted to specific environments. For example, the intensity of consumer participation can be varied, the number of generation and power saving installations, included in the smart energy network, is flexible and a smart grid can be applied as a local mini grid as well as on a large scale. That is, due to several opportunities for re-invention, the implementation of smart grids and smart off-grid solutions under different conditions is facilitated.

As the preceding reasoning showed, despite some challenges, expert judgement did not reveal any major technological barrier to the implementation of smart grids in Mozambique. The vast potentials of re-invention can reduce compatibility problems and facilitate a sustainable implementation of smart grids. However, upgrading of existing infrastructure might come with difficulties due to potential threats to the inter-operability of smart and conventional components. Leapfrogging can circumvent this problem in areas where up to now, no grid infrastructure exists. Mini and off-grid solutions offer additional possibilities to adapt the systems to the specific local conditions.

5.5.2. Economies of density

With a population of only 26 million on an area of about 800,000 square km (CIA 2016), Mozambique possesses a relatively low density of population. Additionally, a large part of the Mozambican population is scattered across the vast country: In 2015, about 67.5% of Mozambicans lived in rural areas (World Bank 2016 b), many of them in very small communities and remote villages (Government of Mozambique 2009). The low density of both, population and energy-intensive industries in many parts of the country, has consequences for the feasibility of grid investments. The more people and companies are concentrated in a certain area, the more strongly economies of density occur. Economies of density are cost reductions, resulting from an increasing density of clients. Typically, higher densities of clients lead to decreasing average costs if supply infrastructure shows a high level of indivisibility.

It is characteristic for grid infrastructure that the capacity of one unit of infrastructure cannot be increased continuously but only by discrete steps – to a certain extent it is *indivisible*. Therefore, usually the capacity of a certain unit of infrastructure is not exploited completely, especially in the case of a low density of clients. For a certain unit of grid infrastructure, fixed costs are constant and independent from the number of clients. At the same time, the share of fixed costs in relation to the variable costs among the total costs is normally very large, since large investments for physical infrastructure are only supplemented by relatively low costs for grid operation (Fritsch 2014). That is, if newly connected clients lead to more current in the grid, very little additional transmission costs arise and fixed-cost degression leads to lower average costs.

Figure 7 illustrates economies of density for an infrastructure, marked by indivisibilities, for a case of high fixed costs for initial investment and low variable costs, corresponding with low and constant marginal costs. The underlying assumptions of figure 7 represent a non-peak load scenario. It can be clearly seen that economies of density lead to decreasing average costs: The higher the number of clients[22] in the infrastructure´s catchment area, the more economically feasible it becomes. For example, the higher number of clients y_2 corresponds with the lower average costs ac_2.

The vast majority of the experts mentions explicitly that the low density of population and the low density of energy-intensive industries in Mozambique is an important barrier for smart grid infrastructure. Consequently, GIZ´s representative and the scientist of Eduardo Mondlane University

---

22   The model assumes equal power-demands among the clients.

say, off-grid solutions could be more viable than grid connections when it comes to rural households. Former research has also reached the conclusion that off-grid electrification can be a feasible option to tackle the problem of low demand bundling in rural African areas (Karekezi 2002).

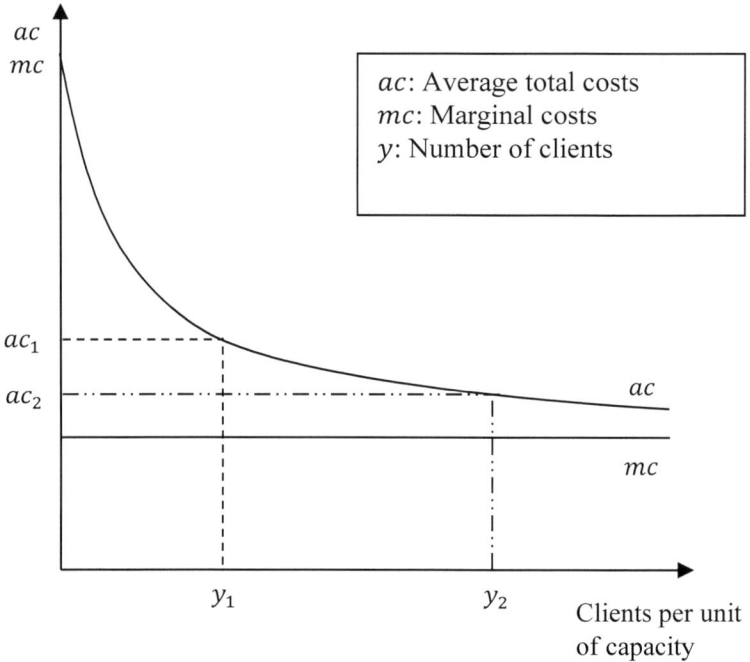

Figure 7: Economies of density.
Source: Own illustration based on Fritsch (2014, 162).

Estimates by the World Bank (2015, 38) show that in Mozambique, even in a low-cost scenario, the costs of a connection vary between US$ 1,400 for a rural customer with need for improvement of the distribution network and US$ 358 for a basic single-phase connection in an already existing network, as usually found in urban areas. Already this basic example illustrates the large relative cost advantage of urban in comparison to rural connections.

Since a lack of economies of density favors alternative solutions to central grid infrastructure, the chance of implementation of smart energy solutions increases for rural areas if they can noticeably improve the performance of off-grid or mini-grid electricity supply. A problem of small grids or off-grid installations is that shortages cannot be balanced by power from other regions. Smart appliances might be a promising option to harmonize demand and supply patterns in off-grid solutions and mini-grids. Thus, energy security in smaller autonomous systems can be improved, the representative of the Ministry of Mineral Resources and Energy says. Further research shall reveal, which smart solutions most effectively meet the prerequisites for electrification in different regions of Mozambique, given the existing drivers and barriers. A closer look at the consequences of economies of density for different electrification options will be taken in chapter eight.

5.5.3. Costs and benefits of a smart energy sector

When the willingness to pay for smart grids was discussed, we already saw that demand depends on the relative price of a smart grid. A smartly interconnected grid is more cost-intensive than a conventional one. In addition to investments for transmission and distribution, smart appliances to monitor and manage the smart grid have to be implemented. Besides these investments into information and communication technology, a fully functioning smart grid would require investments into load flexibility such as storage systems. Furthermore, components of the conventional grid might have to be upgraded to be compatible with the smart grid (EPRI 2011).

Precise numbers for additional costs of a smart grid in Mozambique or similarly structured economies have not yet been revealed and remain a subject to further research. In general, though, the overall benefits of smart grids are expected to outweigh the costs (IEA 2012). For the United States of America, the Electric Power Research Institute, EPRI (2011) estimates a benefit-to-cost-ratio of 2.7 to 6.0 in the next twenty years with additional costs of US$ 338 to 476 billion and additional benefits of US$ 1,294 to 2,028 billion in comparison to conventional grid infrastructure, induced by investments for a fully functioning smart grid. The finding that a smart grid´s benefits outweigh its costs is supported for Great Britain in a study by Ernst & Young (2012).

While transferring these estimates to the Mozambican context comes with severe threats to external validity, these numbers can give an idea of the benefit-to-cost-relations of smart grids. According to EPRI (2011), some of the smart grid´s benefits are: Enhanced reliability, increased productivity, more effective incentive-setting, better management of storage and loads, reduction of interruptions and transmission losses, anticipation of disturbances (self-healing), increases in dynamic efficiency and better demand response. The entrepreneur underlines that due to these advantages of smart grids, *"Smart Grids are necessary in Mozambique. They have an enormous potential."*. It should be kept in mind, though, that as a representative of a private business, the respondent can be assumed to have a high personal interest in new grid investments. Therefore, the respondent´s statement might be influenced by personal interest. Nevertheless, the statement portrays the point of view of a relevant actor in the energy sector´s decision process. If this statement is representative for further private energy companies, a high interest in smart grid investments can be assumed among this important group of stakeholders.

Experts expect the high initial costs of smart grid implementation to be a barrier. The representatives of FUNAE and Geneva Institute of International and Development Studies recall the high costs for infrastructure investments and additional investments in information and communication technology, especially in rural areas. However, experts also see the potentials of smart grids. FUNAE´s representative mentions the significant transmission losses which could be effectively tracked and healed by smart grid solutions. For the productivity of the private sector, instable and fluctuating power supply is a problem, as the entrepreneur underlines. Smart grids´ potential for improved load management addresses this problem. The representative of the Ministry of Mineral Resources and Energy adds that learning-curve effects from pilot projects could further help to improve the benefit-to-cost ratio of smart grid solutions.

The question remains whether it is socially and economically profitable to focus investments on smart grid implementation rather than on small-scale power generation or completely different areas such as primary education or the health system, as the representatives of FUNAE and Geneva Institute of International and Development Studies bring up. Efficient allocation requires investments into the area with the highest relative marginal utility. If smart grid solutions can effectively contribute to a better electricity supply in Mozambique, the relative marginal utility of smart grid investments can be assumed as relatively high. The electricity sector is one of the major bottlenecks of economic development in Mozambique and an improved energy supply is expected to produce positive spillovers to other sectors, such as education and health. This point of view is supported by the representative of the Ministry of Mineral Resources and Energy.

Experts unanimously agree, that generally there is a market for smart energy solutions in Mozambique. However, most of them emphasize that the success of smart energy in Mozambique depends on the specific implementation. From the renewable energy scientist´s point of view, in many regions of the country, the implementation of a central smart grid is economically not feasible. Instead, the respondent points out that smart appliances can bring about large benefits for decentral, isolated grids by enhancing their reliability. In many cases, consumption of clients, connected to mini- or off-grid systems is not time-critical. Many electric devices, like phones and batteries for home use are charged at plug stations, shared by the whole community. This bundled energy demand is quite common in rural areas in Mozambique. This common use of one power source could be a promising area to implement small scale demand response technologies. Staying with the example, phones at a common plug station could automatically be charged when power is available. As soon as power becomes short, the station is cut off the supply system to secure an equalized load and secure grid stability. Such small-scale demand response solutions could be supported by smart technologies that monitor demand- and supply-patterns and manage the loads.

These examples show, how modifications of basic smart grid technology can contribute to meet specific local needs. That is, smart grids offer several opportunities for re-invention which facilitates the implementation of this technology under different conditions. A re-invented smart system, like described above, can be beneficial for small communities and improve the cost-benefit-ratio. A re-invented technology that addresses local specificities reduces costly mistakes in handling and encourages the use of the innovation (Rogers 2003, 185).

Where no power supply infrastructure exists at all, leapfrogging can be economically feasible. Here, a better benefit-to-cost-ratio is possible if conventional technologies are leapfrogged in order to directly start with the newest technology, exploiting its potentials for more efficiency. For large grid infrastructure, according to the representatives of FUNAE and Geneva Institute of International and Development Studies, leapfrogging is the most realistic option for a smart grid in Mozambique. In contrast to this, upgrading the already existing grid is assessed to be rather unlikely. In Mozambique, the main goal of the relevant actors is grid extension and not digitalization of the existing grid. That is, digitalization can come as a side effect: If the grid is extended anyway, new power lines can be smart ones.

Summarizing the findings, smart grids are assumed to be economically feasible in Mozambique but it crucially depends on how they are implemented. Upgrading of existing grid infrastructure is difficult. Consequently, leapfrogging is a viable option: Grid extensions and new off-grid projects can directly benefit from smart technologies. To estimate a precise cost-benefit ratio, further econometric research is needed. The more precise the cost-benefit analysis is, the better potential investors can minimize risks and estimate the expected results of a smart grid investment.

5.5.4. Grid management

While the variable *maintenance* could not unambiguously be identified as a barrier in the quantitative analysis, the majority of the experts in the qualitative interviews clearly classifies a lack of maintenance in the Mozambican electricity sector as an important impediment to smart grid implementation. GIZ´s representative calls a lack of maintenance even one of the biggest economic problems of the country. Findings of the World Bank (2015, 10) support the view that EDM´s level of maintenance is not sufficient to sustain the current quality of transmission.

Besides the lack of financial resources, a high level of centralization was revealed as one of the major reasons for poor maintenance. Both EDM and the relevant authorities are characterized by a very centralized decision-making and concentration of staff at few parts in the country. EDM´s staff criticizes that EDM only holds four regional authorities, responsible for maintenance. Accompanied by a deficient transport infrastructure, this centralization brings about a lack of information about maintenance necessities, high transport costs for personnel and delays of operational procedures. As an effect, EDM´s representatives describe that the operation of the grid in many parts of the country is happening in an isolated way by local staff without regular checks by central administration. According to them, information and communication technology could be a solution to this problem. If a software existed that monitors and visualizes problems in the grid, maintenance could be significantly improved, they say. Thus, EDM would have real-time knowledge and could immediately react to problems.

Maintenance and operation of smart grids requires distinct knowledge and capabilities. Most experts describe a lack of qualified staff as an important problem for grid management. The problem that formally qualified staff is available but needs additional training due to a lack of specialized knowledge (see ch. 5.2.4) also holds for operation and maintenance, EDM's staff, FUNAE's representative and both scientists say. Even if staff is available, its work is significantly hindered by a lack of equipment and tools, EDM's staff specifies. According to EDM's representatives, funds for equipment and tools are too small or misused.

The lack of competition in the electricity market also affects grid management. According to Eduardo Mondlane University's professor of renewable energy, more pressure from competitors would drive EDM to improve the quality of maintenance and operation. Furthermore, the respondent expresses his experience, that private companies usually have higher incentives to invest in grid management than EDM. One reason might be that private companies have to carry all entrepreneurial risks coming with insufficient grid management, while EDM is backed by the government. Thus, EDM has incentives to show opportunistic behavior since it can exploit the advantages arising from governmental backup while the government cannot entirely observe EDM's behavior. This form of moral hazard originating from biased incentives caused by governmental protection of the incumbent monopoly can be another reason for the lack of grid management (Hölmstrom 1979).

A further important barrier to maintenance can be seen in the relevant actors' priorities. The Mozambican government's ambitious goals to increase access to electricity comes with large efforts for new power lines. Hence, funds are channeled to grid extension but maintenance left with a lack of resources. Besides, there might also simply exist a loose culture of maintenance among the relevant actors (Bugaje 2006). The representative of Eduardo Mondlane University shares his experience that companies in Mozambique do not value maintenance. The entrepreneur adds, better maintenance in Mozambique would require a higher level of involvement in decision-making of local communities. For example, locals could help to identify problems in the grid or even help to fix them. However, this point of view is challenged by EDM's staff who says that one of the drivers to maintenance is that EDM organizes maintenance in-house without any sub-contracts. Further research should show, under which conditions maintenance of the Mozambican power grid can be improved.

5.6. Governance and stakeholders

5.6.1. Goals and political performance

The introduction of smart energy can be significantly facilitated or imped-ed by political goals, priorities and especially the performance of political actors. For goals and priorities, the findings allow for a rather optimistic assessment regarding electrification in general and smart grids in particu-lar. Already in its 2009 energy strategy, the Mozambican government out-lined the integrated extension of glass fiber and power lines as one of its central goals for the development of the electricity sector. Thus, grids shall serve simultaneously for power, telecommunications and data transport. The goal is to increase the economic feasibility of the investments and en-able a better cooperation of the two technologies (Mozambican Govern-ment 2009).

Representatives of Eduardo Mondlane University and the Ministry of Mineral Resources and Energy agree that electrification is one of the gov-ernment´s top-priorities. As already stated before, the government wants to increase access to the electricity grid to 50% of the households until 2023 (see ch. 2.1). GIZ´s representative specifies, though, that the gov-ernment preferred grid-based, central electrification over decentral and off-grid electrification.

Taking a look at EDM, according to its representatives, the utility has a very open policy regarding digitalization and new technologies. However, FUNAE´s representative points out that even if the responsible entities were willing to use the benefits of digitalization, they would be hindered by an alarming lack of knowledge about information and communication technologies.

Besides goals and priorities, what strongly influences economic deci-sions is the actual governance of a country. Regarding the performance of political actors, expert-assessment reveals severe threats. A current event with a high potential to scare off investors and which raises questions re-garding the integrity of the Mozambican government, is the appearance of a large hidden debt, found by international creditors in 2016. The Mozam-bican government had guaranteed loans of probably around US$ 1.5 bil-lion (10% of the Mozambican 2014-GDP (UN 2016 d)) for public compa-nies but did not inform its international creditors such as the International Monetary Fund (IMF 2016). The disclosure of these hidden debts caused a loud reaction by the national and international press as well as by rating agencies. Severe deteriorations of Mozambique´s credibility as a harbor for investments can be assumed.

As a reaction to the deteriorated financial credibility of the Mozambican government, significant amounts of aid funds and budgetary support were withdrawn, Mozambique´s credit-ranking was decreased as a reaction to the disclosure of the hidden depts from B- to CCC by Standard & Poor´s (S&P) and from B- to CC for Fitch. The withdrawal of capital from Mozambique in reaction to this discussion also lead to a severe drop of the Metical which deteriorated from 26.95 MZN/USD on 10th March 2012 to 78.56 MZN/USD on 3rd October 2016 (Sampablo et al 2017, 8). The representative of the Ministry of Mineral Resources and Energy ranks the disclosure of the hidden debts on the same level as the violent conflict (see next chapter), being a major current barrier to foreign investments in Mozambique.

Besides the current lack of political integrity, experts criticize that Mozambican governance is characterized by structural problems. FUNAE´s representative sees severe shortcomings in the performance of political actors. According to the respondent, what holds back investments is a lack of political willingness to support economic development. The respondent complains about a lack of communication and a lack of vision of how electrification goals can be achieved. Representatives from the private sector and Eduardo Mondlane University criticize that although the government officially claims to be willing to support private sector engagement in the electricity sector, it fails to implement incentives for private companies. The entrepreneur even challenges the view, that the government really wants more private companies in the Mozambican power market. According to the respondent, the government only wants to protect its public companies.

Briefly concluding, although the government´s goals and priorities favor electrification and digitalization of the electricity sector, expert judgement indicates that severe shortcomings regarding the performance of political actors harm the quality of governance. Furthermore, major political problems like regular violent tensions and the disclosure of hidden debts have adverse effects on Mozambique´s credibility and its attractiveness for investments.

5.6.2. Political and violent conflict

In the preceding chapters, violent conflicts were already mentioned as a barrier to economic development in Mozambique. From time to time, Mozambique faces tensions between the government party FRELIMO and the armed forces of the opposition party RENAMO. The tensions and their meaning for the country can only be fully understood if one keeps in mind Mozambique´s violent history which still influences the political relations of the present. The remainder of this chapter will therefore portray the current tensions between FRELIMO and RENAMO and present an overview of the conflict´s historical background.

As chapter 0 already showed, experts, interviewed for this study, classify the political conflict in Mozambique as the most important political risk to economic development in general. Since the civil war which lasted from 1977 to 1992 (UN 1992), conflicts between the former enemies FRELIMO and RENAMO still flare up from time to time.

In 2013 and 2016, news about several attacks of RENAMO fighters against civilians and acts of sabotage have brought back old tensions to the country (Wessels 2016). RENAMO blames FRELIMO to constantly oppress members of the opposition. It names arrests and disappearances of several RENAMO officials – which it says, were unjustified and arranged by FRELIMO – as reasons for its return to bases in the vast bushlands of central Mozambique from where it executes its attacks. RENAMO´s major goal in the conflict is to gain political power in the provinces, it claims to have won in the provincial elections and requests to nominate those provinces´ governors (Fabricius 2016).

Until now, violent tensions have not reached the dimension of a new civil war – also because regular mediation talks between representatives of the government and RENAMO are taking place (Fabricius 2016). However, as long as the deeply rooted political tensions remain, risk-averse investors will probably leave out economic ventures in Mozambique. Furthermore, if the current violent tensions continue or even intensify, political priorities will probably further shift from economic development to the fights.

Five of the experts (see chapter 5.2) state that the violent conflict holds back foreign investments and harms economic dynamics. GIZ´s representative adds that prognosis for the future, regarding economic developments in the electricity sector, is difficult since it is unclear how the conflict will evolve. The representative of the Ministry of Mineral Resources and Energy also underlines, that the conflict directly affects the power transmission network. The respondent explains that in some areas, purposely destroyed roads and a high threat of attacks make maintenance and construction impossible. The respondent continues, that grid infrastructure is highly under risk to be a target of sabotage.

What is the background of the conflict between FRELIMO and RENAMO? Unfortunately, scientific publications about the origins of the conflict and about the Mozambican civil war from 1977 to 1992 are extremely rare. The few existing ones are mainly based on primary sources such as interviews with the central figures of the conflict who in some cases dominate Mozambican politics until today.

The understanding of FRELIMO´s and RENAMO´s hostilities mainly starts with Mozambique´s struggle for independence. Under its first leader, Eduardo Mondlane, the Mozambican liberation front FRELIMO, founded to achieve Mozambique´s complete independence from the Portuguese colonizers, declared the "general armed insurrection of the Mozambican people against Portuguese colonialism for the total and complete gain of Mozambique´s independence" on September 25 of 1964. At that day, the violent fight for Mozambique´s independence started. FRELIMO´s invasion was operated from its headquarters in Tanzania (Cabrita 2000, 29).

Already at that time, FRELIMO´s leader, although closely aligned to the United States of America and further Western countries, tolerated financial and military support for its struggle from both the Soviet Union (USSR) and China. Resulting from this support, China and the USSR were well respected among FRELIMO members (Cabrita 2000, 28).

In 1969, during the independence war, the FRELIMO leader Mondlane was assassinated with a parcel bomb. Under his predecessor Samora Machel, FRELIMO fostered its ties with the USSR and turned its back on the Western sphere and, too, on China. For the following decades, FRELIMO remained closely aligned with the communist bloc. Samora Machel intentionally used Marxism-Leninism as an ideological foundation for his revolutionary struggle (Meredith 2011, 312).

After years of war against FRELIMO, the Portuguese forces saw major political disturbances in their home country. In 1974 the Carnation Revolution overthrew the fascist dictatorship in Portugal and paved the way for a democratic republic. In effect, the Portuguese armed forces in Mozambique and their colonial administration were severely disoriented and weakened (Meredith 2011, 311). The weakening of the colonial forces contributed to Portugal´s surrender in Mozambique.

Shortly after the regime change in Portugal, FRELIMO and the Portuguese government signed the Lusaka Accord on September 7 of 1974 which restored independence and regulated the transfer of power to FRELIMO (Cabrita 2000, 4). On June 25 of 1975, the former Portuguese colony became officially independent as the People´s Republic of Mozambique, ruled by FRELIMO which turned itself into a state party. Democratic elections were not part of the transition plan (Cabrita 2000, 5). Instead, FRELIMO imposed an authoritarian regime and claimed to be "the sole legitimate representative of the Mozambican people" (Meredith 2011, 311). The FRELIMO government made plantations and businesses national property, introduced central economic planning and collective agricultural production (Merdith 2011, 312). Eventually the former mass movement turned into a "restrictive party of the Marxist-Leninist elite" (Emerson 2014, 25).

Resistance against FRELIMO´s communist agenda, its authoritarian rule and its inefficient economic politics rose among parts of the Mozambican population right after independence. This discontent was a central factor for the outbreak of Mozambique´s civil war (Meredith 2011, 312 and 611).

At that time, there was also another player involved which contributed to tensions in Mozambique. Parallel to Mozambique´s struggle for independence, the Zimbabwe African National Liberation Army (ZANLA) fought against white minority rule in Rhodesia. FRELIMO tolerated ZANLA bases on Mozambican territory which entailed frequent operations of Rhodesian troops against ZANLA in Mozambique (Emerson 2014, 30). These attacks of Rhodesian units against ZANLA in Mozambique, starting from 1977, are seen as the beginning of the civil war because the operations are closely connected to the emergence of RENAMO.

During one such raid of the Rhodesian troops on Mozambican territory, the Rhodesians freed André Matsangaice, a former FRELIMO official, from a re-education camp, established by the Mozambican government. Matsangaice was given military training and installed as the leader of a new, artificial Mozambican resistance movement, which had been founded by the Rhodesian secret service shortly before (Cabrita 2000, 146 f.). The secret service intended to form a small group of partisans to destabilize the antagonized and anti-colonialist FRELIMO-ruled neighboring country (Emerson 2014, 44). This would-be resistance movement became later known as Resistência Nacional Mocambicana or RENAMO (Cabrita 2000, 146 f.). It has become evident that RENAMO's origins cannot be considered a bottom-up movement from the population but rather a puppet organization, intentionally created in a top-down process for obvious political interests of Rhodesia.

The Rhodesians initially intended to keep the resistance organization small enough such that it would remain under control of the Rhodesian secret service and its military (Emerson 2014, 42). However, the size of RENAMO's army and the number of its supporters rose quickly. While it only put 76 fighters on the ground by September 1977, this number rose to 914 by the end of 1978 (Emerson 2014, 44).

One central instrument for RENAMO's success was the Rhodesia-based radio station Voz da África livre ("Voice of the Free Africa") (Cabrita 2000, 139 f., Emerson 2014, 35). The radio station identified itself with the resistance's goals and supported RENAMO with its program and with air time for members of the Mozambican resistance movement. Furthermore, Voz da África livre's program addressed the shortcomings of the FRELIMO government (Emerson 2014, 36). In the first years of RENAMO's operations, Voz da África livre made the resistance appear more powerful and more organized than it actually was. Consequently, ever more Mozambicans began to arrive at RENAMO's bases in Rhodesia to join the struggle for resistance (Cabrita 2000, 139 f.).

Initially, RENAMO did not follow the goal to seize power in Mozambique but merely to create conditions for democratic elections and democratic order. To achieve this goal, RENAMO chose military resistance as its path of preference. The corresponding political process was supposed to be left to externally acquired politicians (Cabrita 2000, 143). From the beginning of the civil war, attacks on civilian and military vehicles, sabotage of roads and demolition of central supply infrastructure such as power transmission lines were a core element of RENAMO's guerilla strategy (Cabrita 2000, 154). Throughout the war, RENAMO became feared for its extreme brutality including exemplary massacres and mutilations as well as the recruitment of child soldiers (Meredith 2011, 611).

In 1979, RENAMO started to fully invade Mozambique and to establish itself permanently in the central Mozambican Gorongosa region (Cabrita 2000, 155). To push forward quickly, Rhodesia's air force supported RENAMO with supplies (Emerson 2014, 44). In the same year, RENAMO's first leader Matsangaice was killed in combat. He was replaced by Afonso Dhlakama (Emerson 2014, 58) who was the RENAMO leader until his death on May 3, 2018 (Mandlate 2018).

Under Dhlakama, RENAMO stepped away from being solely a military movement and gave itself a political agenda. Through diplomatic missions in Lisbon, RENAMO formed strong links with parties and organizations in Western countries. In exchange with the Western political actors, Dhlakama and his combatants realized that Western supporters were only willing to invest in a movement, that makes clear which political goals it wants to achieve. In reaction to that, RENAMO gave itself a political manifesto which claimed for democratic elections, a free market economy and an abolishment of the communist ideology in the Mozambican government (Cabrita 2000, 246 f.).

After the Zimbabwean independence, the winds for RENAMO changed. The new Zimbabwean government supported the FRELIMO government and the RENAMO elements still located in former Rhodesia, now Zimbabwe, including the Voz da África livre team, had to relocate to South Africa (Cabrita 2000, 157). In many ways, the former support for RENAMO from Rhodesia was replaced by South Africa´s apartheid-regime. Like Rhodesia before, South Africa wanted to destabilize the antagonized FRELIMO regime in its neighboring country (Emerson 2014, 71 f.).

Despite attempts for peace talks, the civil war escalated in the mid-1980s. To the Mozambican government, a military solution of the conflict was the only option. The Zimbabwean National Army supported the FRELIMO government with its airforce that flew attacks against RENAMO´s bases (Cabrita 2000, 235 f.).

In 1986, Samora Machel, FRELIMO´s leader and the first Mozambican president after independence, died in a plane crash. Joaquim Chissano was assigned new party leader and in Mozambique´s one-party system automatically became head of state. Under Chissano, the FRELIMO government agreed to a political solution of the conflict in 1989. The Catholic church mediated the peace talks which took place in Rome (Cabrita 2000, 269 f.).

Already during the peace talks, FRELIMO announced the introduction of democratic elections and a multiparty system (Ansprenger 1997, 99). Furthermore, it changed the country´s name from People´s Republic of Mozambique to Republic of Mozambique. A single, non-partisan army was introduced which was supposed to include RENAMO´s fighters. Thus, central demands of RENAMO were fulfilled during the peace talks

(Cabrita 2000, 270). Finally, a peace treaty was signed on October 4 of 1992. Officially, RENAMO transformed itself into a political party with the goal to participate in democratic elections (Meredith 2011, 612).

At the heights of the civil war, around 100.000 people were under arms (Emerson 2014, 24). Among the devastating results of the 16-year civil war are more than one million people killed by the early 1990s and five million displaced out of a population of back then 18 million people. The impoverishing effects were not less serious: After the war, 90 per cent of the Mozambicans lived below the poverty line and 60 per cent in absolute poverty (Meredith 2011, 611). Furthermore, the country started with a supply infrastructure, especially roads and power lines, which were either destroyed or severely harmed due to a lack of maintenance.

Since the signing of the peace treaty of Rome, every five years presidential elections and elections for the parliament took place, though accompanied by allegations of severe manipulations and fraud. In 1998, the opposition, led by RENAMO, boycotted the local elections. RENAMO claimed the conditions of the elections were unfair (Cabrita 2000, 271). Furthermore, RENAMO until today accuses FRELIMO of an incomplete integration of and discrimination against former RENAMO soldiers in the joined armed forces. As already described above, still today RENAMO and FRELIMO are in a constant debate regarding RENAMO's demand to receive the right to govern the provinces where it won the majority of the votes in the national elections. Further issues of discussion are the decentralization of political power in Mozambique, the disarmament of the RENAMO militia and RENAMO's accusation of discrimination against its members by the government (Jornal Notícias 2017).

Despite unquestionable successes of the peace process, such as an existing multiparty system and democratic elections, the peace between the former enemies remains unstable. Starting in 2013, the political clashes between FRELIMO and RENAMO have led to a return of the still existing military arm of RENAMO to the bushland in central Mozambique, performing attacks on buses and acts of sabotage against infrastructure (Fabricius 2016).

Currently, the conflict only continues on the political stage since the conflict parties find themselves in a long-term cease-fire, announced on 4 May 2017 by RENAMO-leader Dhlakama. Nevertheless, the expert's judgements, mentioned above and the deeply rooted historical origins of the conflict indicate that the tensions between RENAMO and FRELIMO remain a strong barrier to economic and political development. A conflict, as long and as deeply planted into the people´s minds as this one, can be expected to persist with strong path dependencies. Peacebuilding is therefore a challenging and time-intensive task, especially after RENAMO-leader Dhlakama died on 3 May 2018 (Mandlate 2018). Which path RENAMO will take after the death of its long-time leader is still unclear. It has to be expected that the conflict between RENAMO and FRELIMO will not be resolved in the close future but remain a burden to Mozambique´s development and consequently also for the development of Mozambique´s energy sector for a long time.

5.6.3. Public institutions

Experts criticize a strong rent-seeking behavior among the staff of public institutions in Mozambique. The entrepreneur describes that administrative processes are infused by a system of bribery. GIZ´s representative specifies that it is a matter of bribe how far one gets in procedures like for example approval processes for business. Thus, initial costs to start a business are increased due to corruption and barriers for market entry are strengthened: Bribes for market entry are sunk costs for incumbent companies but not for potential competitors. Thus, competition in the electricity market can be held back due to the high prevalence of corruption. Besides, if public staff is mainly focusing on rent-seeking instead of economic development, resources are used without productive output (Tullock 1967).

Since, according to expert assessment, Mozambican officials are often easily seduced by corruption, the high abundance of natural resources can pose a severe threat to economic development. Experiences in many countries show that the abundance of natural resources can have a negative impact on the quality of governance and public institutions. Besides economic reasons like the appreciation of the national currency due to the export of natural resources, political effects such as an increased vulnerability to corruption, a strong concentration on extractive industries instead of manufacturing and services and a stronger short-term orientation due to high rents in the present can lead into a so-called resource trap (Sachs, Warner 1999, Venables 2016). Bearing in mind the existing prevalence of rent-seeking behavior, paired with a vast abundance of natural resources, Mozambique is presumably under risk to be affected by such politically induced resource traps.

Besides the system of bribery, experts from GIZ, FUNAE and Eduardo Mondlane University criticize a missing qualification of public servants, a structural lack of staff, lack of qualification and a lack of clarity about responsibilities, leading to cross-over working and to an extension of administrative processes. However, experts also point out that the Mozambican government is working to fight corruption and to make institutions more efficient. For example, the representative of the Ministry of Mineral Resources and Energy stresses that the government implemented an inter-ministerial council which controls public institutions and denounces officials if corrupt behavior is detected. This assessment is to be treated with caution, though, as the respondent represents the criticized governmental institutions.

With the support of international development agencies, the Mozambican government implemented so called "one-stop shops" and continues to do so in order to speed up administrative processes. As GIZ´s representative describes, one-stop-shops aim to bundle all administrative processes that are required to start a company. Thus, representatives from GIZ and Eduardo Mondlane University summarize, many administrative procedures for companies could be sped up and bureaucracy could be reduced. Nevertheless, the representatives of GIZ and the private renewables company say that the level of bureaucracy in Mozambique remains enormous compared to most industrialized countries. The only way to circumvent the sometimes purposely complicated processes, is a bribe, they say.

It is worth to stress that EDM´s staff did not bring up any criticism against public institutions. In contrast, one of the respondents even explicitly states that the respondent´s department (projects and financing) never experiences serious problems with authorities. A possible inference might be that due to the strong bonds between EDM and the Mozambican state, processes are easier for EDM than for competitors.

EDM´s staff alongside with the representatives of FUNAE and Geneva Institute of International and Development Studies states that the excessive centralization of decision-making in both EDM and Mozambican public institutions leads to a lack of information and knowledge about the situation in many parts of the country and results in inefficient planning and slower electrification.

Centralization leads into a situation in which power and control are concentrated in the hands of only a few individuals. In effect, a strongly centralized structure usually lowers an organization´s innovativeness. The problem in a centralized organization is that the range of new ideas taken into consideration for implementation is restricted by the low number of strong leaders who dominate the organization. Especially problems on the operational level cannot be observed as well by leaders as by the operational staff itself and are therefore under risk to remain unaddressed. On the other hand, if one of the strong leaders is convinced of an innovation, the implementation of the innovation is usually easier in centralized and hierarchical organizations since the decision of a highly-positioned executive is more powerful. That is, centralization impedes the early adoption of an innovation but – once adopted – bolsters the implementation of the innovation (Rogers 2003, 412).

It has to be considered that the Mozambican government has its reasons for a strong concentration of decision-making in the capital Maputo and in provincial capitals, controlled by the government: Decentralization means more power for RENAMO, the party representing the majority of the population in some central and northern provinces of the country. In its provinces, RENAMO staffs many of the local and regional authorities. Since RENAMO regularly poses threats to separate the provinces under its control from the rest of the country, decentralization can be a risk to the unity of the country.

According to the experts, many public institutions in Mozambique are furthermore characterized by an impeding conservative culture. The entrepreneur points out that a strong suspicion against private companies is prevalent among public servants which holds back competition and innovation in the energy market. The representatives of GIZ and the Ministry of Mineral Resources and Energy add that a huge effort is needed to convince officials of the advantages of new technological opportunities like renewables or smart solutions.

In a brief summary, in spite of first steps to improve efficiency and effectiveness of public institutions, structural shortcomings remain. First of all, corruption holds back economic development. Furthermore, strong centralization, a lack of qualified staff, bureaucracy and a conservative corporate culture are prevalent among Mozambican public institutions.

5.6.4. Regulatory framework

In the World banks´ 2016 "Ease of doing business index" which evaluates countries´ regulatory performance and the quality of administrative processes, Mozambique became 133 out of 189 in the overall ranking after rank 128 in 2015 (World Bank 2016 c). This ranking goes together with the judgements expressed by representatives of FUNAE and the private company. The respondents claim that Mozambique is characterized by a lack of legal security. According to them, investors cannot fully rely on Mozambican authorities and have to calculate a risk of failing investments due to legal and political insecurities.

In the 2016 sub-categories of the index, specifically relevant for the regulatory agencies, Mozambique is ranked 31 in "dealing with construction permits", 66 in "resolving insolvency", 99 in "protecting minority investors", 105 in "registering property", 124 in "starting a business" and 184 in "enforcing contracts" (World Bank 2016 c). These rankings reflect that Mozambican regulatory agencies show a relatively good performance in some areas but remain with significant shortcomings in an international comparison, especially regarding legal security. However, if we only look at Sub-Saharan African countries, Mozambique is on position 14 out of 44. That is, in a competition of African economies, Mozambique´s result reflects some advantages (Ibid.).

In the meantime, Mozambique´s rankings in the "Ease of doing business index" have not changed for a better. In the overall ranking of 2018, Mozambique drops 5 ranks in comparison to 2016 to rank 138. Mozambique significantly deteriorates in the categories "dealing with construction permits" (from 31 in 2016 to 56 in 2018), "resolving insolvency" (from 66 to 75), "protecting minority investors" (from 99 to 138) and "starting a business" (124 to 137). The country remains stable for the categories "enforcing contracts" (rank 184 in both years) and "registering property" (rank 104 in 2018 after 105 in 2016) (World Bank 2018 b). In the inner Sub-Saharan comparison, Mozambique also remains relatively stable with a rank of 16, only two ranks below its position in 2016. (World Bank 2018 c).

It is reasonable to assume, that the instable political situation in Mozambique in the years from 2016 to 2018 with a boiling violent conflict and the disclosure of hidden debts, did not pass by without influence on Mozambique´s position in the Ease of doing business index. Mozambique´s unfavorable position and development in this index indicate that effective political reform to improve the position of private business in Mozambique was not undertaken since 2016.

In the interviews, respondents especially criticize a missing scheme of incentives for companies in the Mozambican electricity sector. Respondents from FUNAE, the private sector, GIZ and the Ministry of Mineral

Resources and Energy state that an effective incentive scheme (like efficient subsidy schemes) for new and renewable technologies and small and medium enterprises is missing (see also 8.4). FUNAE´s representative concludes that regulation of the Mozambican electricity market clearly favors the incumbent monopoly EDM. Furthermore, the respondent doubts the existence of any effective regulation of the Mozambican energy sector: *"There is no regulator, a legal framework is not really there, legal security is not really there."*.

The shortcomings in the Mozambican regulatory framework affect the prospects of smart energy in this country. Legal insecurity, costly barriers to market entry and a lack of (guaranteed) revenues can keep potential smart grid investors away from the Mozambican electricity market. However, in a Sub-Saharan comparison, Mozambique offers relatively good regulatory conditions for the introduction of new technologies.

5.6.5. International cooperation and development assistance

The impact of donors and international cooperation was clearly identified as a driver in the empirical analysis. Mozambique´s economy and government strongly depend on development aid. As already analyzed extensively in chapter 5.3, development aid is a significant source of financial flows into the country. Consequently, aid has significant economic power in Mozambique.

Financial aid, technical support and consultancy by donor agencies can support economic development and foster smart energy implementation if it is among donor´s priorities. The more efficient donors perform, the greater is the support. A lack of coordination or a lack of expertise among donors can increase transaction costs and harm the outcomes. Transaction costs are especially increased, says EDM´s staff member, responsible for grid operation and maintenance, due to donors´ conditionality. According to the respondent, the fact that many donors still attach conditions to their support, complicates the implementation of projects in the electricity sector. Examples of such conditions are certain requirements of governance or making it compulsory to use material from the donor´s country of origin. Since EDM cooperates with a large number of partners, conditionality eventually leads to a mixture of brands and different technological designs that sometimes are not compatible, the respondent states: *"We have many different brands. It´s the conditionality. It makes our job diffi-*

cult.". Thus, need for coordination was higher and maintenance costs increased, the respondent says. The manager´s evaluation is generalized by the representatives of the private company and Geneva Institute of International and Development Studies. These respondents criticize that donors were too little demand-oriented but rather focused on their own interests and their own ideas for projects instead of following the population´s real needs.

At the same time, the entrepreneur continues, a long-term perspective and a sensitivity for the capacities of the local population was regularly missing among donors. For example, the representative of the private sector illustrates that in the energy sector, donor projects often applied expensive and complicated energy systems which cannot be maintained nor operated by the local partners due to a lack of funds and qualification. Thus, capacity building and independent continuation of the projects was made impossible. Consequently, a dependency on the donor who implemented the projects can be created.

Besides creating new dependencies, donor involvement can also hinder self-generated initiative, since aid inflows and support from donor agencies bring about strong incentives to focus on more convenient rent-seeking instead of self-responsible and sometimes risky economic action. Thus, aid can become a curse (Djankov et al. 2008) for economic development. The entrepreneur supports this view, stating that the involvement of donors regularly reduces the self-initiative of the private sector and locals. Furthermore, the respondent states, donor agencies crowded out private companies, since – in contrast to the private sector – donor agencies were typically characterized by a lack of cost sensitivity. Another critical point about donors, the entrepreneur brings up, is that in his experience, whenever his company participated in projects funded by donor agencies, it was confronted with a level of corruption among donors, similar to the Mozambican authorities. That is, if this experience proves typical, a system of development aid not only induces rent-seeking among the recipients but also on the donors´ side.

In general, however, the positive aspects of donor involvement dominate in the experts´ assessments. Except for the scientists, respondents from all institutions under analysis brought up that development cooperation was a driver to smart energy implementation, not only due to the bare capacities – finances and expertise – but also due to a high expected interest in smart energy projects (GIZ) and the high priority of electrification among donors (Ministry of Mineral Resources and Energy). The representative of the Ministry of Mineral Resources and Energy specifies that donors were already involved in projects which aim to combine power generation and transmission with smart technologies. Furthermore, this respondent alongside with GIZ´s representative expects donors to have a high ability and willingness to pay for smart grids (see also chapter 5.3).

Besides development cooperation, south-to-south partnerships can improve economic development in Mozambique. Especially within a cooperation of like-minded countries, spillovers, mutual benefits through trade and technology diffusion as well as an exchange of experiences can enhance the chance of introducing new technologies. Mozambique is a member of the 15-country South African Development Community (SADC), an important actor for cross-border development and political cooperation. In article 5 of the SADC Treaty, it aims to "promote development, transfer and mastery of technology" (SADC 2015). Furthermore, Mozambique is one of the 54 member-states of the African Union which according to article 2 of its charter also aims to "coordinate and harmonize" the economic, scientific and technological cooperation of its members (OAU 1963, AU 2012).

Since the economic and political framework conditions are similar in most southern African countries, it can be inferred more easily from a SADC or African Union country´s experience to the Mozambican context than from an economically, politically and culturally different country. Being located right next to South Africa, Mozambique could benefit from its economically strong neighbor. South Africa is the second largest economy in sub-Saharan Africa (IMF 2018), displays a high demand for electricity and shares a cross-border grid network with Mozambique. However, beneficial trade between Mozambique and South Africa requires fair trade agreements which do not only benefit the economically stronger partner.

Cooperation between South Africa and Mozambique can support the introduction of smart grids, since South Africa has already collected some experience in the implementation of intelligent grid systems. South Africa´s National Energy Institute launched the South African Smart Grid Initiative (SASGI) in 2008, aiming to promote the implementation of smart grids in South Africa´s energy sector (RSA 2008, SASGI 2016).

It is questionable whether the goals to diffuse technologies in the SADC-region are pursued effectively. Interviewing the experts, potentials to improve this cooperation become evident. The representatives of GIZ and the Ministry of Mineral Resources and Energy agree that in the last decades, systematic joint initiatives, aiming to bolster electrification were not taking place on a high level – neither in the African Union nor in the SADC-network. However, the chief of department points out that inner-African cooperation regarding electrification has increased in the last years. It is also a goal of the Mozambican government to increase cooperation in the SADC on electrification issues (Government of Mozambique 2009). Even if political actors did not effectively pursue cooperation for electrification on an inter-governmental level, the continuing economic integration of the southern African economies would favor the diffusion of new technologies and the exploitation of comparative advantages (Ricardo 1817) in the electricity sectors.

It has been clearly stated that development aid, the involvement of donor agencies and international cooperation are expected to be an important driver to electrification and to the introduction of new technologies like smart energy in Mozambique. Despite some problems like crowding out private initiative, the creation of dependencies and an increased need for coordination, donors can probably effectively support electrification and the implementation of smart grids. The economic relevance of aid in Mozambique is high and smart grids are expected to attract donors' interest.

5.6.6. Acceptance and stakeholders

To what extent and how quickly a smart grid infrastructure diffuses through Mozambique, depends not only on how attractive it appears to investors and suppliers but also on how well it is accepted by consumers and other relevant stakeholders in society. Smart grid implementation can only be successful in the long run if it is aligned with the specific Mozambican demands.

Consumers might be reluctant to purchasing costly smart meters; obligatory rollout might face acceptance problems. However, precise harmonization of demand and supply by a smart grid requires a high market penetration of smart meters to monitor consumption profiles (Schreiber et al. 2015). Except for the potential rejection of smart technologies in the case of an obligatory financial participation of customers, experts do not expect any acceptance problems regarding the necessary infrastructure – grid network and smart appliances.

Indeed, FUNAE´s representative states that in Mozambique, it was much easier to realize large infrastructure projects than in most economically more advanced countries since protest against infrastructure of civil society is relatively rare. Also, projects with a significant environmental impact or even the relocation of residents to make room for infrastructure are "no big problems", he says. As long as there was financial compensation, the local population was quite cooperative. It should be considered, though, that the Mozambican government applies quite robust measures to keep the voice of civil society quiet. The government´s typical reaction to protest ranges from cutting governmental subsidies for the organizations concerned, defamation and violent response to disappearances and deportations of activists (Human Rights Watch 2019).

Looking at electrification as a specific kind of infrastructure installation, empirical findings indicate high levels of genuine appreciation and acceptance. Experts do not think that authoritarian governmental measures are even necessary to enforce acceptance but rather describe an intrinsic motivation among the Mozambican population to support the electrification of their country. Generally, many Mozambicans were very open-minded to electrification infrastructure, representatives of EDM say. Giving reasons for this high level of acceptance, they state that there is a very high demand for a better electricity supply and that people understand that electrification requires the extension of the grid infrastructure. GIZ´s representative specifies that many people in Mozambique were very interested in new technologies and supported their implementation, also observable in the in the booming smart phone market. The representatives of GIZ and the renewables company agree that acceptance problems regarding infrastructure can be further avoided if the local population is involved in planning and implementation, benefits from the infrastructure (e.g. grid connection) and is properly informed about the necessity.

Besides the application of infrastructure, data security and privacy are strong concerns in most discussions about the introduction of smart technologies. In Mozambique, however, EDM´s staff and the representative of the Ministry of Mineral Resources and Energy do not expect any relevant reservations to smart grids due to threats to privacy. They point out that data security is not a priority concern in the Mozambican society. As long as data collection can contribute to a better energy supply, they say, society would show understanding and no acceptance problems were to be expected. *"There will be no problems [with acceptance]"*, as one of the respondents, representing EDM, states.

The representative of the Ministry of Mineral Resources and Energy argues that EDM already collects a lot of data about its customers and never faces any acceptance problems. In Mozambique, many electricity consumers use a pre-payment technology: The consumer purchases a certain amount of electricity in advance, types in a code on the meter´s interface and receives the corresponding amount of energy. EDM uses these meters to collect information about consumption patterns.

The experts´ judgements indicate that in contrast to societies in some industrial countries, the Mozambican society does not display major concerns about data security and privacy. Therefore, the necessary collection of data – an important concern about smart grids in Western countries – is not expected to be relevantly challenged by stakeholders in Mozambique. Furthermore, no or only little protest against infrastructure projects is expected, especially in the power sector. A high acceptance of smart technologies and electricity infrastructure is therefore considered a driver to smart energy implementation in Mozambique.

What is the reason for the phenomenon, described above, that most Mozambicans apparently do not show significant concerns regarding data protection problems and the environmental impact of grid infrastructure? Why is the value they put on these issues apparently smaller than in most Western, industrialized countries, where threats to data privacy and destruction of the environment are typically accompanied by strong civil protest? Ronald Inglehart´s theory on value change (1977) offers a well-established framework for the explanation of differences in values between countries with different economic and political backgrounds. The theoretical reasoning and the empirical findings, Inglehart´s theory is based on, do not explicitly incorporate data privacy as variable. Nevertheless, for differences in the appreciation of data privacy among populations, this theory offers a possible explanation as data privacy can be assumed to be ranked among post-materialist needs (Bennet 1992).

Inglehart´s theory assumes that depending on the experiences in their pre-adult years, people form certain hierarchies of values and maintain them during their lives. For example, someone who experiences a strong scarcity of food and/or a lack of physical security in his or her pre-adult years would value the abundance of food and security stronger during his or her whole life than someone who always had enough to eat and generally felt safe (Inglehart 1977, 23). Thus, values are reason and result of a social, economic and political environment (Kersting 1996, 57).

For Western countries, relatively robust empirical evidence exists that formerly rather materialist societies become steadily more post-materialist if they experience economic prosperity and physical safety (Inglehart 1977, 208). While materialists value materialist goods such as income, security and assets, post materialists rather focus on immaterial goods like participation, freedom and self-fulfillment (Inglehart 1977).

According to this theory, a society like the Mozambican one which faces severe poverty, a high prevalence of unfulfilled basic needs and severe threats to physical security would be expected to be strongly materialist. Knutson (1972, 28) argues that poor and unsafe people neither have the energy nor the skills to deal with societal concerns that do not directly affect them. In a study of several countries, including Kenya and Ivory Coast, which analyses the interconnections of urban poverty and political participation, Kersting and Sperberg (2003) concluded that for the urban poor, political participation is impeded by their special situation. According to the authors, "They [the urban poor] use their resources first and foremost in the economic and social sphere. There is relatively little time for politics within their prevailing risk-minimising strategies." (Kersting, Sperberg 2003, 179).

In the absence of post-materialist attitudes, it would accordingly not be expected for a country like Mozambique that concerns with the environment and with data privacy would emerge. It follows that in the light of this theory, the absence of acceptance problems for smart grid infrastructure in Mozambique originates from poverty and permanent violent conflicts that impede the development of post materialist attitudes.

Inglehart explains post-materialist attitudes in societies that have experienced economic prosperity and peace for a longer period of time with the *scarcity hypothesis* and the *socialization hypothesis.*

Scarcity hypothesis: This hypothesis assumes that people put the highest value in their most pressing needs (Inglehart 2008, 131). The scarcity hypothesis is based on the theory of diminishing marginal utility: It assumes that the utility of the next marginal unit of the consumed commodity decreases, the more one consumes this commodity (Gossen 1854, 4 f.) Hence, if the scarcity of material commodities decreases – that is if more food, social security etc. are "consumed" – the marginal utility of these commodities decreases, too. Therefore, a further increase of economic wealth and physical security comes with ever lower increases in utility. Consequently, further improvement of economic conditions and security is assigned a lower priority than before and a re-orientation to new, post-materialist values emerges (Inglehart, Flanagan 1987). In effect, the satisfaction of sustenance-needs leads to a socio-political dissatisfaction and people start to address shortcomings in post-materialist needs (Inglehart 1977, 147-148). Post-materialists urge for belonging, esteem, aesthetic and intellectual satisfaction (Inglehart 2008, 131).

Socialization hypothesis: This hypothesis assumes that values do not immediately adjust to changes in wealth and physical security. In contrast, one´s experience in the pre-adult years shapes one´s values. Thus, value change occurs when a generation replaces another generation if the new generation grew up in different economic and political conditions (Inglehart 1977). This assumption is based on solid evidence from social sciences, according to which one´s basic personality is formed in the pre-adult years and basically fixed once people reach adulthood (Rokeach 1969, 1973, Dalton 1977, Inglehart 1977).

The question remains, in which way a change of values among the members of a society is reflected in political attitudes. It is especially interesting for the purpose of this study, what attitudes on electricity transmission infrastructure and privacy protection go together with a shift towards post materialist attitudes and values.

Inglehart´s studies show that post-materialist individuals are more likely to be left-liberal in political attitudes (Inglehart 1977, 61). They place a high value in individual freedom and rather reject submission to authorities as submission would mean a constraint to individual self-fulfillment (Inglehart 2008, 140). Berry (1999) argues that post-materialist values encourage groups of people to demand civil rights and civil liberties from corporate or institutional power.

Liberal democratic theory calls for a private domain where people can act without interference of government or comparable institutions (Bennet 1992). In line with the findings, mentioned above, it becomes evident that data privacy concerns are rather to be expected among post-materialist individuals than among materialists. Consequently, a strong materialist orientation due to the difficult conditions in most Mozambicans´ formative years can be an explanation, why potential concerns about data protection do not raise acceptance problems for smart energy in Mozambique.

The intergenerational change as an effect of different settings during different generations´ formative years is not only a change of values but also a change of *skills* (Inglehart 1977, 5). The skills acquired in the process of change towards post-materialist attitudes are skills necessary to fulfill the new post-materialist needs, like competences required for active political participation. As data from empirical analysis indicates, the change of values and the change in skills have similar or the same origins: Safety and health, better education, shift of priorities and access to mass media. Under these circumstances, more and more time can be spent on education instead of sustaining basic needs and competences such as to express one´s self, information research or acquiring specialized knowledge can be trained (Inglehart 1977, 296).

Furthermore, people who actively engage in political processes acquire skills "on the job" that help them to increase their engagement´s impact. In the process of participation and active engagement, people develop a sub-jective political competence. That is, the individuals *perceive* themselves as more politically competent in a way to use their possibilities for politi-cal influence to affect governmental decisions – for example by articulat-ing their views in discussions, as groups or non-governmental organiza-tions. They get to know the rules, habits and channels to influence politi-cal decision making.

Obviously, a perceived competence does not necessarily imply that the individual under consideration possesses actual political competences and the capacity to influence administrative processes. However, a certain amount of *"spillovers"* (Almond, Verba 1963, 171) can be expected, since interest in political affairs will probably also lead to acquiring actual com-petences in the field (Almond, Verba 1963).

The development of political skills narrows the gap between the elites and the masses (Inglehart 1977, 296): Passivity and bare following the elites without questioning is abandoned in favor of active participation. In effect, newly acquired skills enable to lift political participation to a higher threshold which is desired due to the new post-materialist values (Ingle-hart 1977, 300).

According to the analysis above, poverty and physical insecurity do not only keep society focused on materialist values but also hinder the devel-opment of skills for political participation. To be a real barrier to smart grid implementation, a lack of acceptance would have to exist and be clearly articulated. That is, not only the society´s values would have to as-sign a high priority to data protection and to the absence on environmental impact but society would also have to possess the necessary skills to artic-ulate these values. Since both – the change in values and the development of skills – are impeded by the economic and political circumstances in Mozambique, most probably acceptance problems will not arise for the implementation of smart energy in Mozambique.

As the theoretical reasoning has shown, if Inglehart´s theory is transfer-able to the Mozambican context, the absence of acceptance problems can be a result of the absence of post-materialist needs and the skills to achieve them, resulting from the experience of economic problems, pov-erty and/or physical insecurity in most Mozambicans´ formative years. Although Inglehart´s theory appears coherent with the situation, found in Mozambique, one may raise the question: Is Inglehart´s theory – with its empirical foundations based on findings from Western industrialized countries – transferable to the Mozambican society?

If and how quickly value change takes place, strongly depends on the specific cultural, historical and political framework in a country (Inglehart 1977). For example, societies that were severely harmed in their privacy in the past might react more sensitively to privacy violations. Nevertheless, according to Inglehart´s findings, nations do not diverge systematically when it comes to the origins of value change (Inglehart 1977, 94). That is, no systematic difference shall be observed in the process of value devel-opment if countries face comparable framework conditions.

Although the countries observed by Inglehart are all industrialized countries they differ significantly in their economic and cultural back-ground as well as in the period, they became relatively wealthy. Consider-ing that despite these differences, all the countries observed experienced similar value changes, it can be carefully assumed that Inglehart´s findings are also transferable to a country like Mozambique. Inglehart himself sup-ports this view when he argues that the projected value change would also occur in non-Western countries if the necessary conditions were prevalent (Inglehart 2008, 137).

Assuming, Inglehart´s findings are transferable, Mozambique has to be considered a materialist society as was already mentioned above. Alt-hough no empirical data exists for Mozambique specifically, findings from quite similar countries support this inference. For Zimbabwe, South Africa and some other South-Eastern African countries, strong evidence for a rather materialist value system exists (Inglehart 2008, 138).

Also, data from the organization *World Value Survey* supports the find-ing that societies in East-African countries are rather materialistic. World Value Survey is an international scientific organization, founded by Ingle-hart which has collected time series data about human values from all over the world from 1981 until today. Even in the very comprehensive World Value Survey data, Mozambique is not among the countries under analy-sis. Neither is Angola which could serve for approximations to the situation in Mozambique, given the close cultural and historical links between the two countries. However, in the publications of the World Value Survey, Mozambique´s neighboring countries like Zimbabwe, Zambia and South Africa and further East African countries like Kenya are clustered around survival and traditional value structures. These countries are far from being categorized as countries with mainly secular-rational and self-expression values (World Values Survey 2017). The categorization of East-African countries as countries with a strong prevalence of survival and traditional value systems in their societies is an indication of rather materialist values in Mozambique since the cultural background of these countries can be assumed to be similar to the situation in Mozambique.

Given, these inferences are valid and the Mozambican society is characterized by rather materialist value structures, it is reasonable to assume that the Mozambican society would prefer an improvement of power distribution over data privacy or environmental protection. The scarcity of electric energy as a basic need implies a higher marginal utility of electrification than the fulfillment of post material needs like privacy and environmental protection.

Nevertheless, the finding, that East-African societies are characterized by a rather materialist value structure does not imply that there are no post-materialist tendencies at all. For instance, empirical findings by *Kersting* (1994) revealed a strong demand for more participation among the individuals of a marginal settlement in Zimbabwe. Indeed, participation was valued most important in the sample of respondents with 57.3% wishing for more participation, followed by rather materialist needs, such as order in the society (56.2%) and fighting inflation (55.7%) (Kersting 1994, 149). These findings indicate, that even in poor and insecure East African regions, post-materialist values are not entirely absent.

Furthermore, there are some developments that might induce a value change towards post-materialist values among many Mozambicans. Besides the strong economic growth, technological innovations like mass media and the internet, changes in the occupational structure towards the tertiary sector, and the expansion of education can be factors that drive the fulfillment of basic needs and lead to a re-orientation in the value structure. If this value change takes place in Mozambique´s future, the strong acceptance of smart grid infrastructure can be challenged by a stronger valuation of privacy and environmental issues. However, the perception of physical security – the second necessary condition for value change towards post materialism – is still far away for most Mozambicans. The permanent conflicts between government and rebels will probably impede the development of post material values because the basic need for physical security remains unfulfilled.

Taking all things into consideration, the little concern with privacy violations or environmental impact will probably persist in the close future. Severe acceptance problems for smart energy solutions in general and smart grid infrastructure in particular are therefore not to be expected. To verify this prediction, additional research will be required though, as so far, Mozambique is not included in long term value analysis.[23]

---

23 An example for such a tool which fits the Mozambican context quite well and acquires direct-response data from the population is FrontlineSMS (Frontline 2017) or comparable services. Alternative services to FrontlineSMS with comparable possibilities are listed in Srinivasan 2014, 81.

6. Preliminary conclusions and further steps

The findings of this study clearly indicate that smart grid implementation in Mozambique faces several barriers but also awaits some drivers. According to the results of this study, a severe lack of capital, heavy market power of the governmentally backed energy utility EDM and too low and inflexible power tariffs are strong barriers to smart energy implementation in Mozambique. Not as significant but still relevant barriers are: a missing willingness to pay for smart grid solutions among important stakeholders, high transaction costs and a poor regulatory framework that lacks legal security. Decision-making is characterized by an inefficient centralization and expensive information processing. Expert judgements indicate that willingness to pay for smart solutions is limited by the high initial investment costs, paired with short-term orientation of potential investors who devalue the future benefits of a smart power system. Grid extension in general is impeded not only by a lack of funds but also by a low density of population that makes large infrastructure investments unprofitable and rather favors off-grid solutions. Furthermore, the development of the Mozambican electricity sector is severely hindered by a high prevalence of corruption, problems with the maintenance of infrastructure, a lack of qualified staff, strong barriers to market entry and a constant threat of violent conflicts between the government and RENAMO rebels.

While a costly and technologically challenging smart upgrade of the existing grid infrastructure is regarded as rather unrealistic, experts see leapfrogging as a viable option: Since large parts of the country remain without grid access until today, new grid systems in these regions could directly be smart ones. The results indicate that for new electricity infrastructure in general and smart appliances in particular, a high level of acceptance among the Mozambican population and other stakeholders can be assumed. Further drivers are expected within development cooperation, the strong growth of the Mozambican economy and Mozambique´s participation in international organizations, such as the Southern African Development Community (SADC). Experts highlight the strong financial power of development aid and the generally positive experiences with donor agencies in the energy sector. Additionally, the respondents expect a high interest among donors in a smart electrification and mention that first pilot projects for power supply with smart appliances have already been implemented in Mozambique.

The progressing economic integration of the SADC member states can bolster cross-border electricity trade and technology diffusion. For the process of electrification, smart grids can be a viable solution to tackle many of the Mozambican power sector´s current problems, like regular blackouts, transmission losses, power theft and poor load management. Especially for isolated grid systems, smart energy solutions bring about potentials for an increased stability.

Based on the results which could be produced until here, the second part of this study aims to identify which path of electrification and which smart energy solutions are most suitable for the Mozambican context, given the identified drivers and barriers. Which smart appliances can most effectively use the Mozambique-specific drivers and circumvent the barriers? Can Mozambique copy the smart grid infrastructure of other countries or are tailored solutions necessary? Which energy design, e.g. a national grid network, smart mini-grids or off-grid systems, suits best the Mozambican context? Policy implications from this research can contribute to the relief of energy poverty by enabling a successful implementation of a smarter, safer and cheaper power supply in Mozambique.

7. Options for a smart electrification

Is there a better and smarter electricity supply in Mozambique´s energy future? If so, what will it look like? Will there be an intelligent main grid system, covering the whole widespread country? Will there be a diversified electricity supply with isolated and interconnected smart mini grids? Will large parts of the country depend on off-grid solutions? Or will the barriers to a smart electricity system be too high, such that there will be no substantial improvement nor smart upgrade of power supply in Mozambique? These are the key issues of interest for the second part of this analysis.

To answer these questions, one needs to construct a likely scenario for Mozambique´s future power supply. The leading questions for this analysis are: Which design of electricity supply suits best the specific Mozambican context? Which smart appliances can be applied most effectively in this scenario? Policy implications from this analysis can contribute to improving framework conditions for the relief of energy poverty in Mozambique.

Conducting the research, it has to be considered that the landscape for smart electrification is very diverse in Mozambique. Some areas are served by the central grid, some by decentral solutions. Large areas have no substantial electricity supply at all. These differences create many cases for the potentials of different options for electrification (Tenenbaum et al. 2014, 35).

Very broadly, there are three ways to provide electric energy to currently unserved regions in Mozambique (Sampablo et al. 2017):

-   Extension and improvement of the **main grid**,
-   **Mini grids** which may be completely isolated, interconnected or linked to the main grid,
-   **Off-grid electrification** by standalone systems that supply individual households.

The extension and improvement of the main grid can be summarized as the central approach. The second and the third approach can be described as the decentral track. Each of these approaches has different implications for the usefulness and feasibility of smart technologies. The different approaches are subject to a close description in the following chapters.

7.1. Central track

The conventional track to electrify a region is expanding electricity access by extending a central main grid. This has been the common approach in densely populated, industrialized countries. A central main grid typically consists of high voltage transmission lines and regional distribution networks with lower voltage. Since this approach to electrification is very common and was already described in detail in the first part of this study, the following paragraphs will only present a brief overview of this approach. For more detail check once again chapters two and three.

Electrification by central grid investments is typically organized by only a few or even just one central actor, like a public energy utility. Consequently, the central track for electrification is typically a top-down approach. The prior driver to electrification within this strategy is the extension of the main grid (Tenenbaum et al. 2014, 1). While extending the grid, the operator has to ensure a supplementary increase of generation capacity and grid management such that additional distribution networks are reliably fueled with sufficient load.

An electricity supply that relies on a central main grid requires that all sources of energy generation and all points of consumption are connected to the overall grid network. Managing the interactions of all points of supply and consumption is relatively easy if the number of sources is limited. A conventional power supply system relies mainly on large, centralized generation, such as coal, gas, oil, nuclear energy or large hydro plants. Furthermore, conventional central grid systems are built upon a one-way distribution of power from the plants to the consumers. Consequently, in a conventional, unidirectional power supply system with a low number of centralized generation sources, load management remains at a relatively low level of complexity.

The more energy generation is diversified and decentralized, the more difficult and complex load management in a central grid becomes, especially if the market share of decentral sources increases due to fluctuating renewable energies. Also, increasing consumer participation is challenging for centrally structured grid systems. If consumers adapt their energy consumption to the scarcity of power using intelligent devices, conventional consumption patterns are not reliable any more. Besides, if consumers add energy to the system by installing small scale energy generation, e.g. by generators or solar panels, they replace the one-way energy flow by a two-way flow. In this two-way energy flow, power is transported from the grid to the households but also from the households into the grid. These former bare recipients of energy become so-called "prosumers" (Rathnayaka et al. 2011) – actors consuming and producing at the same time.

It becomes evident that technological progress, diversification and the increasing market share of decentral renewable energies place challenges on conventionally structured central grid systems. To deal with new challenges, such as decentral and fluctuating generation as well as consumer participation and two-way energy flow, enhanced load management, remote monitoring and control as well as sophisticated data analysis were developed. These technologies are key features of a smart grid. However, a smart upgrade is not the only way to address these new challenges for centrally structured grid systems. An alternative is a decentral approach to electrification. The *decentral track* combines decentral production with decentral distribution of electric energy. Those decentral options come with new potentials for the electrification of remote, sparsely populated areas and will be analyzed in the next chapter.

7.2. Decentral track

The decentral track is a bottom-up approach to electrification (Tenenbaum et al. 2014, 20). Several entities, such as private companies, nongovernmental organizations, cooperatives or communities alongside with conventional utilities organize the provision of electric energy. These small power producers provide off-grid systems or operate mini grids which typically supply power for not more than one or a few local communities (ibid.).

In off-grid or mini grid systems, power is typically supplied by small scale power sources, such as small-scale solar panels or wind turbines, diesel generators, mini hydro installations or batteries (Tenenbaum et al. 2014, 1). Decentral power generation can be supplied by only one of these sources or by cogeneration. A typical example for cogeneration is a solar energy installation which is supported by a diesel generator for back-up energy to guarantee a reliable flow of energy, even when solar power runs short during cloudy episodes or at nighttime (Tenenbaum et al. 2014, 21).

While decentralized electrification is a rather unconventional way to bring power to a region, first experience and scientific evidence show that the decentral approach can be successful in providing a basic level of electric energy to a large part of populations, especially in rural Africa and other developing regions of the world (Tenenbaum et al. 2014, 16). Decentral solutions are especially attractive for countries with very limited funds such as Mozambique because they can usually be applied at much lower costs than central grid installations. Still, decentral solutions and especially interconnected mini grids can serve as building blocks for a larger grid system: Decentral electrification allows for step-by-step electrification, such that a larger grid system emerges from the bottom to the top (Blyden, Lee 2006, 1).

The central and the decentral approach do not necessarily exclude each other. Indeed, they can create synergies: While utilities can focus on central grid extension, where it is feasible, remote areas can be electrified by decentral solutions. Therefore, decentral solutions are of special interest for communities which are not expected to be reached by the central grid in the near or medium term (Tenenbaum et al. 2014, 241). Especially rural communities benefit from an electrification strategy in which central and decentral electrification mutually supplement each other. If decentral power supply is available, these communities gain access to electric energy much sooner than if they had been obliged to wait for the central grid to arrive (Tenenbaum et al. 2014, 287). If the main grid arrives at the site of an independent power generation and distribution network, central and decentral infrastructure can be interconnected. Thus, decentral generation and distribution infrastructure can increase the capacity of the overall central grid network (Tenenbaum et al. 2014, 287).

In decentral electrification strategies, electricity supply is shifted from a centralized top-down actor to many local actors. If regional power infrastructure is not owned and operated by some far-away energy utility but by local actors, new possibilities for participation of local communities are created. Especially if local electricity infrastructure is provided by or in cooperation with the local community or it´s representatives, energy supply can be democratized and organized in participative decision-making processes. Chances are high that awareness of ownership among local community members is increased by having a stake in the installation and operation of essential local infrastructure. Ownership in local communities can be further supported by the fact that instead of benefitting some outside investor, the revenues from locally owned power infrastructure remain in the region where the energy is generated. A higher perception of ownership is expected to lead to better maintenance and sustainability of the infrastructure (Tenenbaum et al. 2014, 7).

The market power of the central utility is moderated if more and more decentral actors enter the market. From an economic perspective, less market power is desirable. The reduction of market power typically improves the efficiency of allocation and results in a more competitive market. From a technological perspective, advantages of a higher share of decentral infrastructure in electricity generation and distribution are: curtailment of transmission and distribution losses, greater robustness against extreme weather and less time-intensive deployment (Platt et al. 2012, 187).

On the other hand, loads can be balanced more easily in large grid structures than in small decentral mini grids or off-grid systems. In large grid systems, grid management can typically rely on many different generation sources which can back-up the system if other sources fail to deliver power. Therefore, decentral solutions come with higher requirements regarding load management and balancing of supply and demand. Smart solutions can help to cope with such challenges in decentral power distribution (Tenenbaum et al. 2014, 7).

In very sparsely populated and poor areas, often even small scale solutions cannot recover their costs. Therefore, governmental actors aiming to boost electrification shall take precaution when designing the regulatory framework for decentral solutions. Administrative requirements and subsidy schemes can influence the feasibility of decentral electrification significantly. On the other hand, regulatory agencies shall ensure that basic quality and security standards are met by small power producers and distributors to protect the population (Tenenbaum et al. 2014, 5 f.).

From the point of view of governments, utilities, donors or other actors pursuing electrification, there is a trade-off between central grid extension and the implementation of mini grids or off-grid solutions. If their funds and effort flow into decentral solutions, there will be less resources left for central grid extension. To determine in which cases decentral solutions are the best option, a closer look at the different possibilities for decentral electrification will be offered in the next chapter.

7.2.1. Isolated mini grids

Since the term *mini grid* is used to describe many kinds of power distribution infrastructure, a precise definition of the term is required to start this chapter. Tenenbaum et al. (2014, 36) define an isolated mini grid as "[...] a stand-alone, low-voltage distribution grid that is supplied with electricity from one or more small generators that connect to only the isolated mini grid."

A *micro grid* – a term occasionally used in publications about decentral electrification – is considered a sub-form of a mini grid in this study. Usually, the term micro grid refers to very small-scale mini grids which typically serve fewer than 150 households. Micro grids usually have an output of only up to a few kilowatts, just enough for basic energy consumption such as lighting, radio or cell phone chargers (Tenenbaum et al. 2014, 44).

While larger mini grids are often operated with alternating current (AC), micro grids often distribute direct current (DC) at low voltage. Thus, the direct current which for instance is delivered by solar panels, can directly be fed into the grid and delivered to the households without AC-DC converters needed. Furthermore, direct current can be distributed with less technical effort, thinner and cheaper wiring. Also, the low voltage reduces the risk of fire or safety problems. However, DC-micro grids do not meet the quality standards of regional mini grids or the central grid which transports alternating current to facilitate the transmission of high-voltage current across long distances. Nevertheless, DC-micro grids are a relatively cheap, safe and simple option to satisfy basic energy needs (Tenenbaum et al. 2014, 44).

Experience from countries like Mali – where already in 2014, more than 150 regional mini grids were in operation – show that the typical mini grid in developing countries serves a few hundred customers and consists of a generation capacity of less or slightly more than one megawatt and of a distribution network of a few kilometers (Tenenbaum et al. 2014, 37). Economic feasibility and reliability of mini grids can be enhanced by a diversified power generation and storage. Typically, mini grids are fueled by solar or diesel generators or a combination of both, supported by a battery. If fossil and renewable sources are combined, generation is described as hybrid power production (Tenenbaum et al. 2014, 2). [24]

---

24   For detailed practical examples of mini grids see Tenenbaum et al. (2014, 37).

Isolated mini grids can be a viable option in countries where the national utility is slow in extending the central grid or where incentives to extend the central grid are low (Tenenbaum et al. 2014, 37). Both is the case in Mozambique, as the analysis of drivers and barriers in the first part of this study has shown. Therefore, mini grids might be a promising option for areas in Mozambique which are unsupplied so far.

Recent advancements in smart grid technology are expected to improve the performance of mini grids. Platt et al. (2012, 189) describe some areas where mini grids can be substantially improved by smart technologies:

- Increased reliability: A smart mini grid is capable of sensing system conditions and reconfiguring generation and device operation.
- Increased resilience: Failures and interruptions are reduced by sophisticated monitoring and self-healing, real-time response and forecasting. Furthermore, automatic reconfiguration in the case of interruptions and device failure avoid system collapse and outages.
- Reduction of emissions through improved integration of distributed renewable generation and balancing of fluctuations in energy supply.
- Taken together, these potential improvements reduce losses, transaction costs and failures. Thus, efficiency is improved, and operation costs can be reduced.

Before these technologies were available, it was much more difficult to maintain a stable power supply in mini grids than in the central grid. As already stated before, it is more difficult in a mini grid to cope with small fluctuations or problems with a generation source than in a large grid system which can back up failures of a generator or problems in one grid region with loads from other sources or regions of the grid. Through developments in smart grid management and in the use of renewable energy, loads in mini grids can be diversified and integrated in hybrid systems with different sources. Furthermore, storage capacity can be used and managed more effectively to balance loads and to provide back-up power. Consequently, mini grids have become more reliable, more dynamic in adapting to changing conditions (e.g. lack of sun for solar panels or lack of diesel for generators) and more flexible in dealing with intermittent renewable sources (Platt et al. 2012, 189). Furthermore, several smaller isolated mini grids can be interconnected to form a larger isolated mini grid to increase stability. Eventually, mini grids are an option, worth to be taken into consideration for areas which are not feasible for central grid extension in the closer future.

7.2.2. Connected mini grids

Very basically speaking, a connected mini grid is a mini grid which is linked to the main grid. Such a connected mini grid may be a fully integrated part of the main grid or operate mainly as an isolated mini grid with a supplementary point of common coupling. A point of common coupling is a connection to the main grid which allows for basic interaction between mini grid and main grid like import or export of power (Platt et al. 2012, 189). The first case follows a completely integrated approach. In the second case, the mini grid operates basically as a distinct system which uses a connection to the main grid to balance loads.

The core feature of a connected mini grid is that electricity is transferable between the main grid and the formerly isolated mini grid. Thus, peak demands in the mini grid network can be buffered more easily by back-up power from the main grid. However, this improvement of load management is only possible if the main grid is not excessively unstable. For the client of a formerly isolated but stable mini grid, the connection of the mini grid to a very unstable main grid can come with a deterioration of energy quality.

If mini grids are connected to the main grid, grid coverage in total is increased. Therefore, linking mini grids to a main grid can be described as a bottom-up extension of the main grid. However, in contrast to top-down central grid extension, instead of one central actor, many individual actors – the operators of the mini grid – contribute to grid extension in the bottom-up approach. Thus, a higher level of burden sharing, division of labor, diversification of risks and participation of local communities is achieved in the bottom-up approach than in a top-down grid extension strategy executed by one central utility.

If power sources in a connected mini grid are well managed such that loads in this new section of the central grid network are effectively balanced, the connection of a mini grid can bring additional stability to the main grid. The connected mini grid can be managed like an additional distribution capacity which can absorb loads in the case of surplus energy and it can be treated as an additional generation capacity to feed power into the main grid in case of peak demand (Platt et al. 2012, 193).

When a mini grid is connected, it is an important question how economic and technical connectivity between main grid and mini grid is ensured. There are several models how connection can be organized when the main grid and a mini grid meet:

1. The operator of the mini grid stops its own generation, purchases power from the main grid at wholesale and resells it at retail to its customers. In this case, the operator of the **mini grid becomes a pure distributor** depending on energy from the utility. This option might be viable if the mini grid operator's generation capacities operate very inefficiently and buying power from the utility is more economical (Tenenbaum et al. 2014, 14).

2. The operator of the **mini grid ceases distribution and focusses on production**. As a small power producer, the operator sells electricity at wholesale to the central utility while the utility owns and operates the grid including the former mini grid infrastructure (Tenebaum et al. 2014, 14). For instance, this option can be beneficial for a small power producer that receives an attractive financial compensation for energy fed into the utility's grid.

3. The operator of the **mini grid is active in generation and distribution.** As a regional power utility, the former mini grid operator manages the connected mini grid and its generation facilities but also trades power with the main utility across the point of common coupling. Furthermore, the new regional power utility sells own and bought power to its customers at retail (Tenenbaum et al. 2014, 15). This approach can be attractive for a stable, efficient and resilient mini grid which only uses the connection to the main grid to balance loads and to generate additional revenues from selling surplus energy.

4. The **central utility buys the mini grid** and its generation facilities. A core prerequisite for this option is that the mini grid infrastructure meets the quality standards of the utility (Tenenbaum et al. 2014, 15). Depending on the utility's offer, selling the mini grid can be more attractive than continuing the operation of a connected mini grid.

5. **Abandonment:** In case of insufficient quality or a serious lack of compatibility, the mini grid can be abandoned and replaced by parts of the central grid (Tenenbaum et al. 2014, 15). If this is the likely case in a country, incentives to invest in mini grid infrastructure are lower, especially in regions where the central grid is about to arrive soon.

Each of these options requires specific agreements and payments between the operator of the mini grid and the utility operating the main grid. Especially issues like power purchase, subsidies, compensations for the operation of infrastructure or benefit sharing are to be concluded on. The core question is: What happens when the main grid arrives at the site of a mini grid? Since clear legal rules increase planning security for this case, the interconnection of mini and main grid is an important issue for regulatory agencies.

7.2.3. Off-grid electrification

In rural, unelectrified areas in Mozambique and many other regions in Africa, until today, car batteries, battery torches, kerosene lamps or natural primary energy sources like wood and charcoal are the most important sources of energy services. In Mozambique, most of the energy consumed comes from biomass (Chongo Cuamba 2006, 76). Using these very basic energy sources is often extremely time-consuming (e.g. collecting firewood), costly and harmful to one´s health (e.g. use of charcoal for cooking). Off-grid systems are an alternative to conventional and inconvenient energy sources for areas without access to a power grid.

Off-grid electrification is a general term for many different solutions with different levels of energy quality. Possibilities reach from solar phone charging, small individual diesel generators, solar lanterns and flashlights to portable solar kits and solar systems for individual households or even businesses, schools and health centers (Tenenbaum et al. 2014, 20 ff.). Usually, in off grid-electrification, not electricity as the final product is sold but small-scale installations to generate electricity for domestic consumption (Tenenbaum et al. 2014, 73).

Summarized as standalone off-grid systems, these installations typically come as combinations of one independent power source, several technical devices such as lights, phone chargers, radio, television and sockets, a storage unit and a management-and-control unit. While larger mini grids with a good load management can catch up or even beat the supply quality of the central main grid, standalone off-grid systems have not yet achieved this level of quality. Usually, their capacity is smaller than grid-based distribution networks and they often depend on weather conditions. Nevertheless, standalone systems enable basic energy use like lighting, phone charging, cooling and access to information and media like radio, internet and television among others (SolarWorks 2018 a).

For standalone systems, several solutions exist. Smaller solutions usually consist of a solar panel, a battery, several lightbulbs, an intelligent management unit and a station with sockets, USB portals and cigarette-lighter plugs (SolarWorks 2018 a). The different components are interconnected by wires, distributing direct current.

In figure 8, a very basic scheme of such a solar home system is illustrated. If sun power is available, electricity from the solar panel is directed to the points of consumption: Lightbulbs or several technical devices, plugged into the socket station. Surplus energy is automatically stored in the battery. At night or in times of cloudy weather when the solar panel does not deliver sufficient energy, the battery is discharged, and energy is directed from the storage to the points of consumption. The process is managed by the load management unit which monitors and controls the loads.

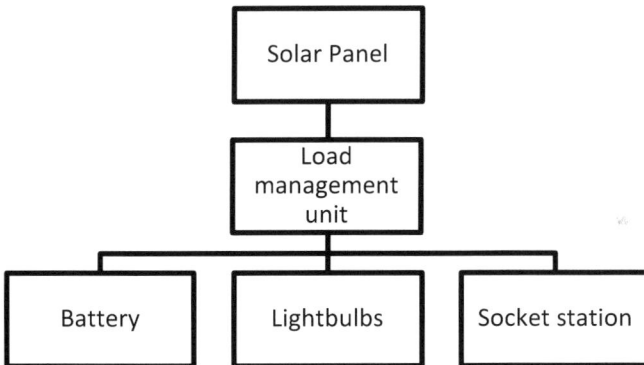

Figure 8: *Scheme of a solar home system.*
Source: Own figure.

Home systems of larger capacity generate even enough energy for relatively consumption-intensive households or even small businesses. They enable to run laptops, printers and modems, to watch an LED-television, use a TV decoder, charge phones and tablets. They usually also come with additional USB ports and sockets (SolarWorks 2018 a).

Several home systems can be connected to each other, such that a small mini grid emerges. Thus, many small generation and storage facilities are included into the network, while each sub-system keeps its own control unit. In effect, generation, distribution and storage can be diversified and reliability can be improved. At the same time, load management stays organized in several controllable and clearly arranged sub-units. Thus, connecting home systems can be an approach to construct a mini grid in a bottom-up approach.

Off-grid solutions offer several possibilities to implement smart technologies. Some possible features of modern standalone systems using advancements in digital technologies are listed in the following paragraphs:

- Solar home systems can automatically switch to alternative connected power sources or storage if solar radiation is insufficient (SolarWorks 2018 a).
- The control unit in home systems avoids overcharging and controls loads (SolarWorks 2018 a).
- A home system´s control unit can collect data on how much power is consumed by with device and when. This data can be sent via mobile communication technology to a central coordinator that processes the data. This data can benefit the provider and the end-user of the system. Specific software solutions in control units enable to change settings remotely, such that no technicians have to be sent out. Consequently, maintenance costs and travelling expenditures are saved and the customers receive assistance immediately (SolarWorks 2018 b). Thus, home systems offer opportunities for remote control, remote response and self-healing. Furthermore, collecting data about consumption patterns allows for a better planning of capacities and system design.
- Using remote control, the provider of the home system knows: Is the system still working? Is the system working optimally? Which customers require an upgrade or an adaptation of the system (e.g. additional installations to increase generated energy)?
- Home systems can be made compatible for pay as you go (PayGo) technologies. PayGo describes a leasing model which allows customers to pay for the home system in daily, weekly or monthly installments. When the full price is paid, the system belongs to the customer. If the customer does not pay, his or her home system is automatically switched-off remotely by the distributor of the home system. Payment for PayGo systems is significantly facilitated by the high penetration of mobile payment technologies in Mozambique. Typical mobile payment systems are MPesa, MKesh or E-Mola. Thus, the users of

home systems can transfer the money across long distances via mobile payment services on their phones (SolarWorks 2018 b).

- If PayGo technology is used, the provider can find out who is paying, when customers are paying and which areas have which ability and willingness to pay. Using this information for strategic planning and for adapting payment requirements, the provider can reduce the downtime of the systems and decrease default rates (SolarWorks 2018 b).

A home system is not only useful for households without any access to the grid. A household connected to a very unreliable and intermittent main grid might be willing to supplement grid energy with a home system to stabilize its energy flow. In this case, a home system can be connected to the main grid by simply plugging the control unit into a socket which is nothing different but a very simple connection to the main grid. Thus, the household can be provided with energy from the main grid to charge the system´s battery or to deliver back-up energy when the home system´s solar panels fail to generate sufficient energy. Modern home systems´ control units automatically switch from its own power source to the main grid if additional energy is needed (SolarWorks 2018 b). In effect, the reliability of power supply can be increased by combining a home system with a grid access.

It can be concluded that intelligent control and management units are the essential device for off-grid solutions to provide reliable and stable energy. These units moderate fluctuations, manage loads, enable self-healing, collect and process data, facilitate payments and switch between different power sources and storage. Therefore, it can be stated that digital solutions are an important contribution to the feasibility of decentral options for electrification.

7.3. Central and decentral track – a brief summary

Some of the essential insights from the preceding chapters are summarized in the following paragraphs. While central electrification follows a top-down approach, decentral electrification constructs a generation and distribution network in a bottom-up process. Figure 9 sums up the core differences between the two approaches and names the specific actors and features of the two approaches.

Looking at the electrification of an entire country, these two approaches are not to be seen as alternatives. Indeed, they can supplement each other as decentral solutions can be feasible in regions where grid extension is not. Both approaches, their corresponding actors and components respond differently to the drivers and barriers, derived in the first part of this study. In the following chapters it will be analyzed what these drivers and barriers mean for the different options.

**Approaches to
electrification**

Central track
(top-down)

Typical actors
- National energy
  utility
- Governmental in-
  stitutions
- Highly integrated
  large companies,
  active on all parts
  of the value chain
  (generation,
  transmission, dis-
  tribution, retail)

Components
- Main grid
- High and medium
  voltage lines
- Centralized distri-
  bution and admin-
  istration

Decentral track
(bottom-up)

Typical actors
- Small power pro-
  ducers and distribu-
  tors
- Private companies,
  organizations, local
  communities,
  households, gov-
  ernmental organiza-
  tions among others

Components
- Isolated and highly
  diversified genera-
  tion, storage and dis-
  tribution
- Mini grids and off-
  grid systems

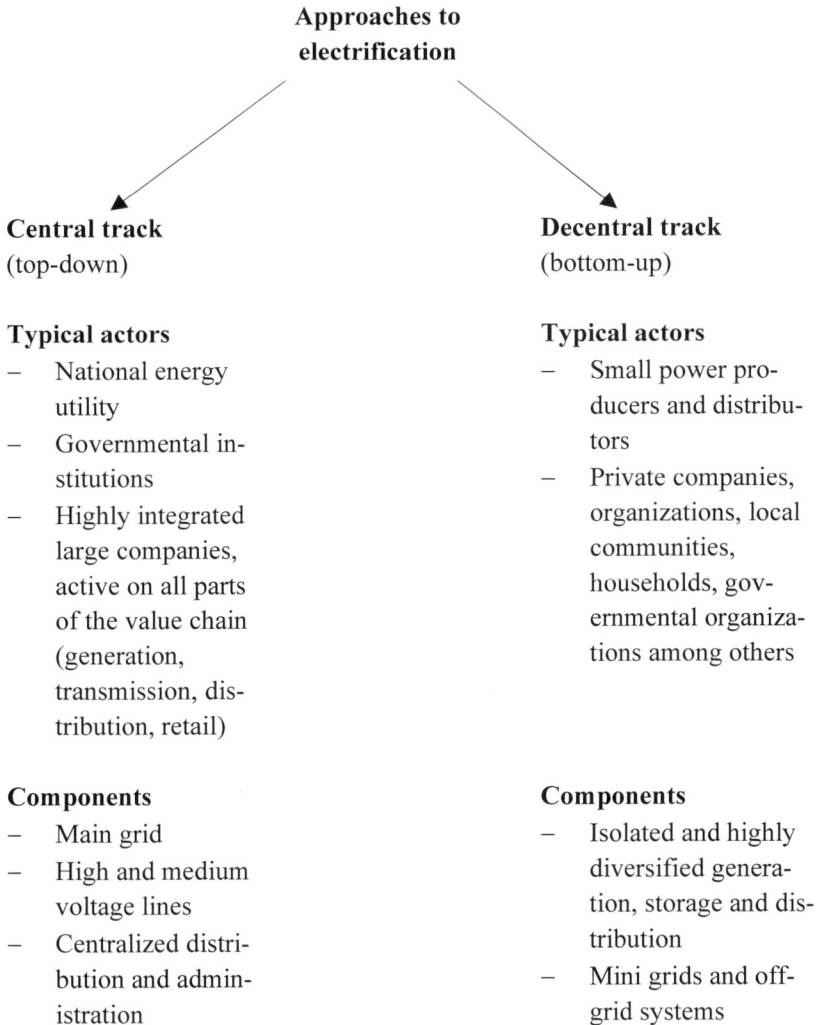

Figure 9: *Approaches to electrification.*
Source: Own illustration.

8. Assessment of Options

In the following chapters, the feasibility of the different options for a smart electrification will be analyzed in order to construct a scenario for the future Mozambican power supply and for the impact of digitalization in the Mozambican energy sector. As the preceding chapter showed, broadly speaking, there are four different options to expand and digitally improve electricity access in Mozambique: Extension and smart upgrade of the central grid, isolated smart mini grids, connected smart mini grids and smart off-grid solutions with each of these options appearing in various forms.

It can be assumed that to what extent a technology is adopted, depends on its relative advantage, compatibility, complexity, trialability and observability as the analysis of innovation diffusion in chapter four showed. As a brief reminder: The relative advantage of an innovation is defined as "the degree to which an innovation is perceived as better than the idea it supersedes" (Rogers 2003, 15). Compatibility is "the degree to which the innovation is consistent with the existing values, past experiences and needs of potential adopters." (Rogers 2003, 15). Complexity is "the degree to which an innovation is perceived as difficult to understand and use" (Rogers 2003, 16). Trialability is "the degree to which an innovation may be experimented on a limited basis" (Rogers 2003, 16). Observability is "the degree to which an innovation is visible" to other potential adopters (Rogers 2003, 16). Taken together, these five factors are assumed to determine an innovation´s rate of adoption (see chapter four) and can therefore be interpreted as an indication of an option´s perceived feasibility.

The impact of the different drivers and barriers on each option shapes their feasibility. While a certain barrier might be very relevant to one option, another option might perfectly circumvent this barrier. While a certain option is pushed significantly by a certain driver, this driver might be completely irrelevant to another option. The purpose of the following chapters is to find out how well the different options for electrification cope with the conditions in the Mozambican power sector, or more precisely, how each of the different options is affected by the drivers and barriers derived previously in this study.

8.1. Methodology of the empirical study

8.1.1. Methodological approach

Due to a structural lack of data and literature about the issue of smart electrification in Mozambique, first-hand information is needed. Accordingly, semi-structured, qualitative expert interviews are applied to analyze the impact of drivers and barriers on the different options for electrification. The empirical research is supplemented by an intensive analysis of the few available literature sources. Furthermore, results from the previous parts of this study complete the analysis.

Expert interviews are regarded as the most suitable tool for the scientific objectives of this study. Profound knowledge of the Mozambican energy sector alongside with knowledge about Mozambican politics, regulation, spatial conditions and economic development are required to make differentiated and reliable statements about the feasibility of options and their potential to benefit from advancements in digital technology.

Qualitative methods are appreciated for their capability to cover complex and differentiated settings (Heinze 2001, Starr 2014). Since the feasibility of options is shaped by a combination of many factors, a very close and differentiated look at the issue of analysis is needed. Due to their open character, qualitative interviews are very sensitive for the specificities and the complex influences of drivers and barriers on the different options for electrification.

Furthermore, qualitative research reveals human perspectives which are very important for the perceived feasibility of different options for electrification. As the analysis of innovation diffusion in chapter four showed, the decision to adopt an innovation depends especially on the subjective perceptions of the innovation by individuals not on objectively verified characteristics (Rogers 2003, 223). That is, besides the given framework conditions, subjective assessments of individuals determine whether or not an option is regarded as feasible or not. Especially subjective assessments by influential actors – such as the decision makers who participated in the interviews for this study – can influence the prospects of an option significantly.[25]

8.1.2. Application of the qualitative interviews

To structure the interviews and to increase comparability, an interview guide is applied. The interview guide contains key questions and additional questions that can be asked to specify the given answers if necessary. [26]

The guide leaves rooms for the respondent to present his or her specific knowledge but confines the conversation to the issue of interest and to the respondent´s expertise. Questions in the guide aim to generate insights about the feasibility of smart options for electrification in Mozambique. Therefore, the impact of drivers and barriers on the options and the options´ potential to benefit from digitalization are covered. Furthermore, the guide contains questions about option-specific aspects of regulation. Regulation affects each option differently and different options require different regulative frameworks. The questions about regulation are regarded as too option-specific to be covered by the general analysis of drivers and barriers in the first part of this study.[27] Therefore, questions about the regulation of the Mozambican energy sector are included into the empirical analysis in the following part.

In qualitative research, it is essential to identify suitable respondents. Within this study, a person is considered a suitable respondent if he or she possesses accessible knowledge about the feasibility of different options for electrification in Mozambique, their potential to benefit from digitalization and their specific regulative requirements. The respondents´ information shall be based on actual experiences or education in the area of interest, not on speculations (Meuser, Nagel 1991). To ensure a broad view and to enhance representativeness, experts should be recruited from different positions of the political-economic process (Mayer 2008, 42). For the purpose of this research, experts were recruited from energy utilities,

---

25   This chapter only describes the main reasons qualitative interviews were chosen as the method for empirical research. For more information about the principles of qualitative research go back to chapter 5.1.

26   The complete interview guide is attached in annex B 2.

27   For more information about the methodological foundations of interview guides as an empirical tool see Annex B and chapter 5.1.

governmental and regulatory agencies, private sector organizations, donor agencies, research and consultancy. More precisely, the following 9 actors were interviewed (in brackets: short form, used in the following evaluation):

-   Head of department, evaluation and planning, FUNAE
-   Manager of projects and financing, EDM, Directorate of Electrification and Projects,
-   Scientific advisor for renewable energies, Ministry of Mineral Resources and Energy of the Republic of Mozambique
-   Technical advisor, Ministry of Mineral Resources and Energy of the Republic of Mozambique
-   Senior consultant for decentral energy, EnDev program (Energising development), donor-funded program to support electrification in Mozambique[28],
-   Senior consultant for solar energy, program EnDev (Energising development),
-   Head of program, development of electricity sector in Mozambique, Deutsche Gesellschaft für Internationale Zusammenarbeit mbH, GIZ,
-   Professor at Eduardo Mondlane University, Maputo, Chair for Physics of the Renewable Energies. Also, this respondent is the chief executive officer of a small-scale company, selling and installing mini grid and off-grid installations based on renewable energy. Additionally, the respondent is the current president of the general assembly of the Mozambican association of renewable energies (AMER 2019).
-   Researcher at Eduardo Mondlane University, Maputo, Department of Engineering. In her research, this respondent focusses on mini grid implementation and renewable energies in Mozambique. Before changing to Mondlane University, the respondent has worked eleven years as a civil servant at the Ministry of Mineral Resources and En-

---

28 Energising Development (EnDev) is a comprehensive multi-donor partnership which pursues the diffusion of sustainable energy technologies that meet the needs of poor households. The program is financed by the governments of six countries, precisely the Netherlands, Germany, Norway, United Kingdom, Switzerland and Sweden. The cooperation agency GIZ represents Germany in the EnDev program (EnDev 2019).

ergy of the Republic of Mozambique, regional directorate of renewable energies in Maputo. [29]

The interviews were realized in spring 2018 and were generally conducted in person and in Portuguese. One exception is the interview with GIZ´s head of program which was conducted via skype and in German. The interviews lasted approximately between 30 and 60 minutes. The interviews were ended when the conversation was not expected to deliver any additional relevant information.

Two of the respondents – EDM´s manager of projects and financing and Eduardo Mondlane University´s professor of renewable energy – were also interviewed in the field research about drivers and barriers for the first part of this study. Due to their broad expertise in the topic of interest, they are also viable respondents for the second field research. Even though the other respondents from the first field research were not interviewed specifically for a second time, their statements are used in the evaluation of the options´ feasibility if parts of their reasoning address this issue. As a reminder, these are the respondents from the first field research, excluding the two respondents who were also interviewed in the second part:

- Manager of grid and transport, EDM, Directorate of Grid and Transport, Department of Power Lines,
- Manager of grid management and maintenance, EDM, Directorate of Grid and Transport, Department of Grid Management,
- In-house consultant of FUNAE, especially in charge of off-grid electrification and mini-hydropower plants
- Head of Department of Renewable Energies, Ministry of Mineral Resources and Energy of the Republic of Mozambique
- Director of MOZITAL, medium-sized Mozambican company in the field of renewable energies, founded in 2006, since than broadly 1900 installations in broadly 120 villages,
- Former consultant of Deutsche Gesellschaft für Internationale Zusammenarbeit mbH (GIZ), expert for electrification, especially renewables and off-grid solutions,
- Professor at Geneva Graduate Institute, Graduate Institute of International and Development Studies.

---

29   A list of all experts is attached in annex C.

In total, 18 in-depth interviews were conducted in the first and second field research. Respondents represent knowledge and experience from various perspectives. Since many of them are involved in political or economic decision-making regarding electrification in Mozambique, their assessments are of high relevance for the diffusion of options, the implementation of digital solutions and regulation of Mozambique´s energy sector.

Unfortunately, so far, there is no distinguished national organization which represents explicitly the interests of the consumers in the Mozambican energy sector, such as a specialized consumer protection agency. It would have been interesting to include assessments of the different options for electrification from a determined consumer protection perspective. However, experience in the interviews showed, that the consumer´s perspective is considered and covered especially by the scientists and by the representatives of development organizations. Donor agencies typically claim to advocate the interests of the consumers as their goal is to benefit the local population and to improve their well-being (GIZ 2018). All scientists explicitly deal with the impact of electrification on consumers in their work.

8.1.3. Data evaluation

To ensure a careful evaluation, the interviews were taped and carefully transcribed in the form of a result protocol.[30] The goals are to find patterns such as common opinions and insightful reasoning (Mayer 2008, Starr 2014). This process is accompanied by an analysis of the statements´ inner logic.

The focus in the analysis rests on determining the feasibility of different smart grid and smart off-grid options in the given Mozambican context. The lead question for the analysis of the respondents´ statements is: How does each option perform under the given drivers and barriers? Additionally, specific regulative requirements for each option are analyzed. Even-

---

30   Transcripts of the interviews are attached in annex D.

tually, policy recommendations are derived from the respondents´ statements.[31]

The impact analysis of the drivers and barriers, based on the interviews and literature research, will take place in the next chapter. There will not be presented a comprehensive description of the drivers and barriers from the first part for a second time. If additional information about some of the factors, influencing the options´ feasibility is needed, it is recommended to go back to chapters 5.2 to 5.6 which contain a comprehensive description of the drivers and barriers.

8.2. Central smart grid

8.2.1. Impact of the barriers

Currently, EDM connects around 120.000 new customers to the central grid each year. At this pace, even until 2030, more than 50% of the Mozambican population will still not have access to the main grid. Initially, though, EDM targeted 50% access until 2023 (see chapter two). According to the Department for International Development, the development agency of Great Britain, for the 50%-goal to be realized, starting from the existing grid infrastructure of 2016, approximately USD 9 billion would have to activated by EDM, the public sector and donors (DFID 2016, 1). However, an essential reason for the slow extension of the grid is a severe lack of funds, as the analysis in the first part of this study has shown.

Reasons for the large lack of capital are EDM´s structural deficits and the low ability to pay of potential customers – especially in rural areas. The representative of the Ministry of Mineral Resources and Energy highlights that in rural areas, people usually do not possess large energy consuming devices, such as fridges or washing machines. Therefore, electricity consumption is low and refinancing the investment and maintenance costs of heavy infrastructure like a central power grid becomes challenging. Consequently, financing gaps have to be closed with external capital, especially from development banks and agencies. Willingness to pay for

---

31   For methodological challenges of qualitative methods in general and semi-structured expert interviews in particular check again chapter 5.1.7.

electrification, grid extension and grid improvement is undoubtedly present among donors. However, donor money can hardly compensate entirely for the immense lack of capital for electricity grid infrastructure in Mozambique.

Besides shortness of capital, the concentration of market power in the Mozambican energy sector can be assumed to affect the pace of grid extension. EDM´s dominating position in the market comes with incentives to exploit market power, for example to charge high tariffs or to keep away potential competitors. Furthermore, market power incentivizes to keep the quantity of distributed electricity low such that EDM can focus on well-paying customers and maintain electricity as a luxury commodity which can be sold at higher prices.

Since EDM´s market power originates from a stable natural monopoly – the grid – EDM is in a comfortable position without actual or potential competition. Concludingly, there are no strong economic incentives for EDM to extend the grid or to improve energy quality if it was not for political goals to improve access to energy in Mozambique.

The price cap regulation is supposed to ensure that even poorer households can afford electricity. From the perspective of grid extension, however, a price cap can have adverse effects for social justice. Due to the lower revenues resulting from the price-cap regulation, EDM suffers from a structural lack of capital for investments. Consequently, funds for connecting new customers to the grid are missing. Thus, the price cap only benefits customers connected to the grid but impedes the enlargement of access to further parts of the population. Especially in rural areas with widely spread customers, grid extension is expensive and can hardly be refinanced with price-capped power tariffs. Consequently, incentives for the utility to extend the grid to rural areas are low. Accordingly, EDM´s manager for projects and financing blames the regulation for EDM´s missing investment capacity. Even after the latest increases, the respondent says, tariffs are still far from being cost reflective.

Besides the revenue problem, EDM´s cost problem impedes grid extension. Additionally, the scarcity of equipment and tools makes maintenance, grid improvement and grid extension difficult and costly say the representatives of the Ministry of Mineral Resources and Energy. Furthermore, EDM has to bear high transport costs and the company´s inner processes are described as inefficient. Real-time information about the functionality of the infrastructure does not exist, digital data collection about grid performance and load management are at very basic levels.

Consequently, transaction costs are high and the potentials of modern error management, grid control and load management remain unused. At the same time, upgrading an existing infrastructure with digital technologies towards the development of a smart grid is extremely costly.

The rigidity of the Mozambican tariff regime is expected to impede the extension of access by a smarter energy supply. GIZ´s head of program criticizes that the absence of flexible tariffs makes smart solutions less viable. Only if tariffs represent the current scarcity of power, smart technologies can use price differences to optimize power supply and to reduce costs. Directing electricity to the customer´s devices when power is relatively cheap and cutting off rather unnecessary energy use in periods of high prices is only possible in a flexible tariff regime. Since this possibility does not exist in Mozambique´s rigid tariff regime, potentials to moderate expensive peak demand and to improve overall load management by using scarcity-reflecting price signals remain unused.

An advanced smart consumer participation as a measure to improve stability and to reduce the necessity of costly peak-load energy is not regarded as viable in Mozambique by the representatives of the Ministry of Mineral Resources and Energy due to the absence of technical capacities and a lack of large energy consuming machines in Mozambican households. Instead of washing machines or dishwashers, households which can afford them, usually employ domestic servants, known as "empregadas" as reminded by Eduardo Mondlane University´s professor of renewable energy.

Generally, of all options, the central grid is affected most by the absence of economies of density in rural areas. A central grid is characterized by large infrastructure which cannot be flexibly adapted in capacity. For example, transmission lines are essential to reach remote areas with dispersed clients and cannot be replaced by less costly infrastructure. However, higher voltage transmission lines are only economically feasible if they connect areas which are relatively densely populated. Only if a certain level of customer bundling is available, the high costs of grid infrastructure can be distributed across sufficient consumers (Szábo et al. 2011, 7). Consequently, the low density of population severely harms the economic viability of central grid extension in Mozambique.

The interviews underline the importance of economies of density. The mini grid scientist from Eduardo Mondlane University calls the dispersion of the population the most important barrier to grid extension. Representatives of GIZ and the Ministry of Mineral Resources and Energy highlight that the absence of economies of density is especially strong in Mozambique as the population is extremely spread in rural areas. According to World Bank statistics, even including the cities, only 37 people per square kilometer lived in Mozambique in 2016. In comparison, Germany had a density of population of 236 people per square kilometer in 2016. Even compared to the Sub-Saharan average of 44 people per square kilometer, Mozambique´s density of population is relatively low (World Bank 2018 d). Consequently, according to the Department for International Development (DFID), off-grid connections in Mozambique only produce 6% of a grid connection´s costs. While one connection with on-grid electrification is estimated at on average USD 3,500, an off-grid connection typically does not cost more than USD 200 (DFID 2016, 1).

It should be mentioned, though, that typically, grid-connections are preferable for most customers as they usually allow for a much higher consumption than lower-tear solutions. Energy-intensive manufacturing, businesses and households with many technical devices usually need grid access to cover their energy demand. However, as the numbers above have shown, grid access comes at much higher costs than simpler alternatives.

The shortcomings of the labor market place another challenge to grid-extension in Mozambique. EDM´s representatives state that it is difficult for EDM to hire qualified staff to manage the challenging organization of grid extension. To understand the technical, financial and regulative framework conditions of grid extension, specialists – engineers, political scientists, technicians and economists – are needed. However, these qualifications are scarce for EDM, also because the public utility cannot offer salaries which compare to payments in the private sector, as the respondents´ answers indicate.

The complex challenge of digitalization in the energy sector is impeded by a lack of employees with know-how and experience in the management of smart technologies, as EDM´s manager for projects and financing and the ministry´s technical advisor say. EDM´s staff is rather qualified for conventional analogue grid management which is based on central generation in large plants. Diversified generation, digital management and alternative approaches to electrification are not part of EDM-staff´s prior expertise, respondents claim in the interviews. Additionally, the first part of this study revealed a lack of initiative among EDM´s employees. Consequently, a lack of motivation and innovativeness appears to be an additional important barrier to grid extension in Mozambique.

Looking at the economic and natural environment, there are some additional severe barriers which make grid extension difficult. Due to the poor state of roads and railways, the transport of staff and equipment to the sites where new distribution infrastructure shall be constructed is challenging, time-consuming and costly. Furthermore, the impact of global warming brings about new problems for grid-based power-supply in Mozambique. Since the Mozambican central grid relies heavily on large hydropower plants, lower water levels in the large Mozambican rivers – a likely effect of global warming – reduce the productivity of the turbines. EDM´s representatives mention that already today, such consequences of climate change create challenges for centralized generation based on hydropower.

Looking at governance-related barriers, the political and violent conflict between FRELIMO and opposition is an important factor that affects grid extension and electricity supply. First of all, grid infrastructure is an attractive target for sabotage and plays a strategic role in violent conflicts (compare chapter 5.6.2). Secondly, the central FRELIMO-led government in Maputo is accused to neglect the central and northern territories where the opposition is strong. Such clientelistic politics (Berg-Schlosser, Kersting 1996, 103) of benefitting first and foremost FRELIMO-affiliated areas can influence the extension of the electricity grid. In Mozambique, specific circumstances exist which create incentives to exploit grid extension to benefit the "own people" instead of directing the grid to the areas with the biggest needs. For example, according to representatives of GIZ and Eduardo Mondlane University, EDM does not issue up-to-date and transparent planning for central grid extension. Instead, planning alternated regularly. Under these non-transparent conditions, clientelistic influence on grid-planning becomes more likely and more difficult to detect.

Thirdly, the political conflict, regularly escalating in violent clashes of government troops and RENAMO-rebels followed by enduring and exhausting peace talks, takes away political attention and financial priority from economic and social policies, including grid extension. Since the extension and management of the central grid is run by public actors, this shift of attention affects central grid infrastructure more than decentral options for electrification which are mainly implemented by actors from the private sector that are not or less involved in the political conflict.

Even without the impact of the political conflict, it is questionable if public institutions, responsible for grid extension and operation, are sufficiently qualified to fulfill Mozambique´s pressing electricity needs. The representative of the donor program EnDev who is responsible for the promotion of solar energy states that grid-based electrification is especially affected by the high prevalence of corruption in public institutions, EDM´s high level of inefficiency and its "old school thinking". As the respondent puts it, EDM thinks that new solutions, such as renewables, are a "children game" and is excessively focused on "Megawatts, gas, oil and large solutions".

EDM´s reluctance towards renewables can impede grid extension because especially in Mozambique with its vast abundance of solar radiation, hydro and wind power, renewables can be a promising option to increase generation and to fuel a larger grid. Once again, it becomes obvious how EDM´s conservative corporative culture can slow down the diffusion of innovations in the Mozambican energy sector. Additionally, EDM´s inefficiently high degree of centralized decision-making, creating high transaction costs, impedes the extension of the grid, as EnDev´s representatives claim.[32]

8.2.2. Impact of the drivers

The financial capacities and the strong interest in electrification among donors can be a possibility to narrow the financing gap of planned grid investments in Mozambique. As already mentioned before, donor money can hardly compensate entirely for the immense lack of capital for grid extension and the implementation of smart technologies, though. Donors engage especially in mini grid and off grid electrification. Grid-based electrification is not necessarily the donors´ top priority especially since poor rural areas – a typical target group of donor agencies – can often not benefit from grid extension. What is more, the latest demonstrations of poor governance – such as the disclosure of the hidden debts and the violent conflicts – might further erode donors´ confidence in governmental actors.

Nevertheless, there are examples for donor engagement in central grid extension in Mozambique. For instance, the Energising Development Program (EnDev) which is funded by several donor organizations, including GIZ, supports the construction of low voltage distribution networks at the outskirts of the central grid financially. Since regional distribution is often

---

32 A table which sums up the impact of the barriers on all different options is presented in chapter 8.6.

underdeveloped while high-voltage transmission lines already exist, EnDev pays a certain fee for every connection, EDM realizes, as EnDev´s representatives point out. The goal of this subsidy is not the extension of the grid in a narrow sense but installing a better distribution network for the so-called last mile from the grid to the points of consumption.

Although impeded by the political conflict and the disclosure of the hidden debts, the growth of the Mozambican economy continues and induces improvements in the agricultural, industrial and services sector. Therefore, demand-driven electrification can be induced by these economic developments and additional funds for grid extension can be accumulated. Electrification and economic development can be mutually reinforcing processes which push grid extension and the use of new technologies. Accordingly, Eduardo Mondlane University´s professor of renewable energy states that especially in areas with industrial production, commercial centers or automated processing of agricultural products, grid extension is viable. Especially the existence of decentral hubs of economic activity can bring the grid also to remote areas.

Economic development in the industrial and services sector typically go along with the growth of cities. The share of Mozambicans, living in urban areas, moved from 5% in the 1960s, passing 25% in 1990, to 32,2% in 2015 (World Bank 2016 b). One inference that is drawn by representatives from Eduardo Mondlane University and the Ministry of Mineral Resources and Energy is that urbanization can be a driver to grid-based electrification by creating economies of density. Urbanization leads to a higher density of population and thus to lower average costs as an effect of a better distribution of indivisible infrastructure costs across a higher number of clients. At the same time, electrification can be a driver to urbanization as better access to energy and resulting economic effects like job creation can be a pull factor for people to move to the cities. That is, urbanization and electrification can also be mutually reinforcing processes.

Digitalization can be another factor which pushes the improvement of grid-based energy supply as the first part of this study showed. Experts clearly urge for the introduction of remote monitoring and remote control in Mozambique´s electricity grid. Six out of the nine experts from the second field research say that remote control is undoubtedly a bottleneck and the top priority for the digitalization of Mozambique´s energy sector. According to FUNAE´s head of department, remote monitoring and remote control can help to detect capacity problems and interruptions. Knowing the problematic spots and origins of failure, maintenance can be improved.

Eventually, emergency situations and blackouts can be avoided. EnDev's solar energy consultant puts it drastically. According to the respondent, EDM's grid is "old and full of losses". Therefore, the respondent claims, the grid management must know exactly what the grid's performance looks like in different areas to react rapidly. The respondent demands a modern central interface at EDM's offices which shows the output of all generation capacities fed into the grid, the loads and the places where energy is taken from or lost in the grid.

A basic prerequisite for remote control technologies to work effectively is the abundance of real time data from the grid. If consumption and payment profiles are not available, capacity planning will not be efficient, criticizes the mini grid researcher at Eduardo Mondlane University. The respondent claims: *"we have to know exactly how much energy the clients are consuming."* The renewable energy scientist adds that remote control, automatic response and remote diagnostics are of special value for grid management in such a widespread country with a poor infrastructure like Mozambique since coordination of maintenance and loads can be significantly facilitated by the use of these technologies.

EDM's manager of projects and financing summarizes the current level of digitalization in grid management and remote control and presents an outlook to the close future: *"We [in Mozambique] are very behind. We have practically no basic data. This level of technology is little advanced in Mozambique. But we are on the way. But we have to be faster."* The respondent continues: *"However, we are not standing still. There are new dispatch centers planned which will manage the grid network using the newest advancements of digitalization. This guarantees the exchange of information, grid management and adaptation of different sub-systems to each other."* Furthermore, according to the respondent, a short time back "nobody in Mozambique" was thinking about demand side management. Now it has become an important issue of interest, the respondent says, especially to manage shortfalls, to adapt to consumption patterns and to organize the exchange of energy. Out of all options, central grid energy supply probably has the highest potential for demand response and consumer participation since the biggest potentials for demand response exist in industrial sites or large commercial and residential buildings with a high and diversified energy consumption. These facilities are typically connected to the main grid.

EDM's manager's statements regarding smart grid management and demand response show, that EDM internally discusses issues of smart grid management. That is, awareness of this technology exists. Speaking in terms of diffusion theory, for many grid management technologies, EDM is in the phase of initialization.

Besides conventional load management, smart grid solutions may help coping with new challenges to centralized distribution networks. The representatives from Eduardo Mondlane University mention that not only decentral but also central electricity supply can benefit from the new possibilities that digitalization offers to integrate intermittent renewable energies. EDM´s current grid management is overstrained with the increasing decentral and alternating generation, feeding into the central grid. To moderate alternations and to coordinate loads from decentral renewable sources, forecasting and automated reaction are necessary (Tenenbaum et al. 2014, 227). Therefore, the representatives of the Ministry of Mineral Resources and Energy state that the increasing proportion of intermittent renewables and progressing diversification of energy generation will make smart management necessary in Mozambique´s central grid. EnDev´s representatives describe that digitalization allows for the centralization of the management and decentralization of production by improving the management´s knowledge about what is happening in the grid in the very moment. Thus, the respondents agree, costs in grid-based electricity supply can be reduced significantly. It becomes evident that renewables can improve grid-based electricity supply but at the same time create serious challenges for conventional centralized distribution networks.

The payment of one´s energy bill in Mozambique has already been facilitated immensely by digital solutions. EDM´s customers can pre-pay for a certain amount of energy with an application (app) on the phone or the computer. This app furthermore makes energy use transparent to the consumer and gives him or her control of his or her energy bill. The app also includes further services, such as paying for television or school fees. According to Eduardo Mondlane University´s professor of renewable energy, this mobile payment system facilitates paying for electricity significantly and saves time: *"Before, we had very long queues for buying energy. But fortunately, this situation changed a lot through the system of distant paying."*

What does the introduction of distant paying mean for central grid-based electricity supply? First of all, the utility of the costumer increases as he or she saves time, traveling costs and inconveniences. Furthermore, from EDM´s perspective, payment collection becomes much easier, also in rural areas. Staff which collects payments in remote areas or decentral

EDM billing stores are not necessary anymore. This also decreases the temptations for corruption since EDM receives its money before it trickles uncontrollably to the pockets of local representatives. Additionally, EDM can collect the data about when how much energy is bought in which area. If this data is used effectively, it can support capacity planning to reduce bottlenecks or to avoid inefficient overcapacities in the grid.

Especially in grid extension, leapfrogging can support digitalization. Correspondingly, GIZ´s head of program confirms that for example improved consumer participation in load management can appear "by surprise" through leapfrogging. Apart from that, international spillovers can take digital solutions into Mozambique´s energy supply. Utilities all over the world are strongly investing in smart meter rollouts to improve demand response and to co-ordinate distributed generation (Hamilton 2012, 398). Thus, technology diffusion among important stakeholders is already in process since several years. The higher the rate of adoption of digital technologies becomes worldwide and especially in neighboring countries[33], the more pressure rests on EDM to follow this example. Thus, spillovers from international diffusion channels can further push the installation of smart technologies in Mozambique. EDM might not be among the innovators or early adopters when it comes to smart metering and smart load management but expert statements indicate that maybe as part of the late majority or as a laggard, EDM will be among the adopters.

As the empirical findings in the first part indicate, a smart improvement of the existing central grid or the construction of new smart power lines is not expected to be rejected or blocked by concerns of data privacy. Instead, the empirical results show that acceptance of new technologies is generally relatively high in Mozambique. Regarding information and communication technologies, the rate of adoption is consequently also relatively high (INE 2014). Therefore, if the other framework conditions are supportive, it can be expected that customers will adopt smart home technologies and accept data collection for load management without serious concerns regarding violations of data privacy. Furthermore, grid extension is not expected to be impeded by opposition against the environmental impact of new power lines.

An important actor to push or impede central grid extension is the government. According to respondents representing GIZ and the Ministry of

---

33   See South Africa´s Smart Grid Initiative (chapter 5.6.5).

Mineral Resources and Energy, the government and EDM clearly prefer central grid electrification over alternative options. The connection of public facilities – such as district administrations, health centers and schools – can drive grid extension, especially since the Mozambican government pursues the connection of such facilities and at the same time controls EDM as the responsible agent for grid extension. It is a typical situation that governmental actors and public utilities favor electrification through main grid extension (Sampablo et al. 2017, 37). However, expert analysis shows a discrepancy between the goals and the capabilities of the government and EDM: While grid extension is one of the public actors' top priorities, it is impeded by shortcomings in governance.

8.2.3. Overall feasibility

The impact analysis of drivers and barriers reveals that central grid electrification is the best solution for urban or sub-urban areas in Mozambique. Regarding capacity and energy quality, the central grid is the most desirable option. However, to be economically viable, central grid distribution requires a sufficient level of consumer bundling at the end of the line. Consequently, relatively densely populated areas with sufficient ability to pay – criteria most likely to be fulfilled in urban or sub-urban areas – are a feasible environment for central grid infrastructure.

However, apart from the agglomerations of population around the cities, Mozambique is a widespread country with a low density of population in most parts of the country. Here, distribution of central grid energy is not feasible. Accordingly, the mini grid researcher states: *"To say 'we will cover the whole country with the on-grid option' would be lying."* Still for decades, GIZ's head of program adds, a relevant part of the Mozambican population will remain without access to the main grid. Besides the lack of economic viability in sparsely populated areas, EDM's insufficient capabilities to achieve a higher connection rate hold back grid extension. The low output of the public utility arises from the various financial and administrative problems within the company and the governance problems in the country.

Regarding digital technologies in the energy sector, Mozambique is certainly not among the innovators or early adopters if one looks at central grid distribution. Institutional conditions with a reluctant utility and governance problems impede the diffusion of these technologies. Nevertheless, first successful adoptions of digital technologies such as the use of mobile payment as well as the planning of new and intelligent dispatch centers show that digitalization has already entered Mozambique's central energy sector. Especially remote monitoring and remote control are seen as very beneficial for load management in the Mozambican central grid. Autonomous load management and automated response are regarded as desirable but very costly and complex in the light of the insufficient financial and operative capabilities of the relevant actors in Mozambique's central grid-based power supply. Consumer participation – or more specifically spoken, demand response – in load management tends to be seen as the least viable out of the available smart grid innovations.

An advantage of a central smart grid this that centralized infrastructure is compatible with the existing experiences of EDM and the government. On the other hand, the complexity of a smart central grid may raise concerns among EDM's responsible decision makers, especially since trialability is low. Even pilot projects cannot eradicate uncertainties entirely such that at some point, a decision has to be made whether or not a nationwide implementation of smart grid systems shall be pursued. At this point, there is no second try any more. That is, comprehensive smart grid implementation in the central grid requires a difficult and momentous decision by the utility. Considering EDM's rather conservative culture, it is questionable if the utility is capable to decide on the comprehensive introduction of smart technologies in its grid under these risks. Consequently, upgrading the existing grid is less likely than using leapfrogging potentials to implement digital technologies in new lines, sub-systems and dispatch centers.

Concludingly, smart electrification by the central grid in Mozambique is generally a feasible option. However, the feasibility depends strongly on the area of the country and on the type of digital innovation. Furthermore, the feasibility of smart technologies differs between existing and new infrastructure. The implementation of digital innovations is easier if leapfrogging is possible.

8.3. Isolated smart mini grids

8.3.1. Impact of the barriers

Although infrastructure costs for mini grids are significantly lower than for central grid-based electrification, mini grids still come with relatively high initial investments costs paired with amortization periods of 25 years or more, as EnDev´s consultant for decentral energy underlines. EDM´s representatives state that a lack of ability or willingness to pay among customers can be fatal for smart mini grid projects since the costs of necessary installations – generation facilities, wiring, batteries – have to be assessed in relation to the low average income of Mozambican households, especially in rural areas. Additionally, within mini grid projects, usually costly load management technology is needed. Representatives from FUNAE and Eduardo Mondlane University state that especially if mini grids are fueled with renewable energies, intelligent load management is essential to moderate the fluctuations.

Therefore, not only for main grid infrastructure but also for mini grids, acquiring sufficient capital is a challenge. EDM is not very interested in alternatives to main grid solutions, FUNAE has limited resources and private investors are kept back by the adverse regulatory framework. Furthermore, small power producers and mini grid operators are still often unlikely to receive attractive loans from commercial banks, mainly due to the lack of experience and a lack of up-front equity capital from mini grid investors.

At the first view, mini grids have the advantage, that in contrast to EDM, mini grid operators do not face the regulated and rigid tariff regime. Therefore, one might infer, compared to EDM, mini grid operators had more possibilities to set cost-reflective tariffs. However, even if a mini grid´s retail price is not directly affected by the national price cap, the price cap regulation can act as a de-facto price ceiling for the mini grid (Tenenbaum et al. 2014, 47). FUNAE´s head of department explains that the regulated EDM tariffs are an orientation for the customers such that they are not willing to accept tariffs, substantially higher than EDM´s ones. This high price sensitivity has already inhibited some mini grids projects in Mozambique, the respondent tells. The communities, usually represented by their traditional leaders, did not accept the proposed tariffs. Consequently, mini grid projects were not realized, and the communities remained unelectrified. The high risk of social discontent regarding tariffs is a serious problem for mini grid operators because – in contrast to the government-backed national utility EDM – private mini grid projects must be economically viable to persist. Therefore, cost-covering tariffs are even more important for mini grid operators than for the national utility.

On a mini grid´s cost side, some more specific disadvantages in relation to the central grid arise due to the smaller size of infrastructure. It is a typical attribute of distribution infrastructure and power generation that their output increases disproportionally to its capacity. That is, if capacity is increased by one unit, output, e.g. energy produced or transmitted, increases by a factor larger than one. Consequently, economies of scale lead to decreasing average costs if output is increased (Sharkey 1982, 59). Typical reasons for economies of scale are effects of specialization, better conditions in procurement, better division of labor, learning effects and physical aspects. For example, in distribution infrastructure, the use of materials typically only increases at approximately two thirds if capacity is doubled. This rule of thumb is known as the "rule of two thirds" in engineering (Fritsch 2011, 161). Due to economies of scale and the further advantages of size, mini grids have to deal with a structural disadvantage in investment costs per unit of infrastructure in relation to the larger main grid.

The total costs of mini grid investments can be further pushed up by necessary improvements of the buildings to which grid infrastructure shall be connected. Especially in rural areas, many Mozambican houses do not have a solid roof. Instead, following traditional building techniques, natural materials such as palmtree leaves or bundles of straw are used to cover the houses. These traditional building techniques are a challenge to constructing a stable mini grid infrastructure. Possibilities to deal with this problem are the application of new metal roofs or tailored grid infrastructure which copes with the challenges of traditionally constructed buildings. Adding metal roofs or similar support to the buildings might be unfavorable, though, as traditional materials often cope much better with local conditions like heat or humidity.

Mini grid investments in Mozambique suffer especially from the lack of the necessary market information and spatial data. The Mozambican government publishes few and often outdated data. Relevant information is typically not available online as open access data but has to be purchased and is difficult to find. For example, in 2014, FUNAE published a renewable energy atlas (FUNAE 2014) for Mozambique which contains valuable information about spatial conditions. For example, the energy atlas covers data about the expected solar radiation at potential sites for power infrastructure investments. The availability of these information is undoubtfully a progress for renewable energy investments in Mozambique. However, the atlas has to be purchased at high costs. If necessary information is not accessible, significant noise is introduced into the planning process. Mini grid investors consequently have to bear high costs to acquire relevant information and they have to make decisions under a high level of uncertainty.

Recently, apart from FUNAE's renewable energy atlas, some additional studies about the conditions for rural electrification in Mozambique alleviated the lack of information for mini grid projects to some extent (Green-Light Consult 2016, DFID 2016, Sampablo et al. 2017). Nevertheless, data availability remains at a low level. The corresponding difficulties constitute a strong hurdle for market entry of mini grid projects (Tenenbaum et al. 2014, 316). Additionally, market entry for mini grid investors is impeded by the complicated application processes for funds and by the long administrative processes which typically occur as non-transparent and infiltrated by corruption.

Even if a potential mini grid operator manages to pass the barriers for market entry, it has to deal with the additional problems which occur, once being part of the Mozambican energy market. According to the respondents in the interviews, for mini grids, especially the lack of planning and legal security has very adverse effects. For example, an investor would probably not want to install a mini grid in an area which EDM has already intended to provide with central grid access. However, according to representatives of GIZ and EnDev, there is no planning security for central grid extension. GIZ's head of program calls it a "huge problem" that grid extension plans alternate regularly and that even the few existing plans are not transparent. To illustrate this point, the respondent mentions that some mini grid projects have already failed completely "because EDM announced in the last second: `We are coming with the central grid'". A lack of planning security also affects financing since banks typically require at least some years of security.

Regarding legal security, GIZ's head of program says: *"Legal security is an issue. If I was an investor, I would not really rely on it."* The reasons for the insufficient legal security are diverse and were comprehensively discussed in chapter 5.6.4. Nevertheless, some mini grid-specific shortcomings in the legal and regulatory framework shall be mentioned here.

In Mozambique, decree 48-2007 regulates the responsibilities, involved parties and technical requirements for electrical installations. This decree also includes general regulations for autonomous generation, sale of electricity and sale to the national grid. These regulations are so unspecific that practically, there is no specialized mini grid regulation in Mozambique. However, regulatory requirements are different for smart mini grids than for central grid-based power supply operated by a public monopoly (Sampablo et al. 2017, 37). Mini grids are typically operated by many individual operators and fueled by a diverse landscape of distributed generation. The lack of specific regulation is even aggravated by the fact that there is not even a working authority which executes the general regulatory framework for mini grids. Experts concur that the regulatory authorities National Council for Electricity (CNELEC) and Energy Regulatory Agency (ARENE) which are assigned by the government to regulate the entire electricity sector are not yet fit for work, especially when it comes to the new regulatory challenges arising from smart mini grids. For example, in order to plan a mini grid project sustainably, one needs to know about connection charges, tariff regulation, property rights, data privacy regulations, prerequisites for market entry and the way how concessions are awarded among others. Due to the absence of a clear legal framework which regulates these issues, several respondents see the regulatory framework in Mozambique as a major problem for smart mini grids.

Another result of the regulative gaps is that relevant actors are not assigned clear responsibilities. Respondents especially criticize a lack of coordination between FUNAE and EDM. Although it contradicts its historical purpose of being exclusively a fund for energy projects without operative competences, FUNAE is the Mozambican pioneer for rural electrification, pushing and implementing alternative approaches such as renewables, mini grids and off-grid solutions. However, lately EDM started expanding its activities in these areas, too. Consequently, EDM and FUNAE do not co-ordinate central grid, mini grid and off-grid electrification effectively, such that delays, frictions and redundancies occur. Thus, unnecessary transaction costs and an unproductive use of resources result. Considering the several barriers arising from regulation and the unsatisfying performance of public institutions in general, GIZ's head of program calls not only regulation but also governance "the main problems" to the implementation of smart mini grids in Mozambique.

Looking at the economic, political and cultural environment for mini grid projects, some further challenges occur. According to Tenenbaum et al. (2014, 25), the necessary know-how to operate mini grids and small power production is usually not available in rural African areas in the first place. However, a sustainable operation of grid and generation infrastructure requires long-term, replicable know-how on the site. External knowledge can only provide for short-term relief.

Despite first experiences with mini grids, implemented by FUNAE (e.g. 72 diesel-powered mini grids already in 2014), new projects still face a lack of qualified staff (Sampablo et al. 2017, 35). Especially experience with renewables-powered mini grids is low as several respondents say in the interviews. Furthermore, experience with the use of information and communication technology in energy projects is scarce.

Besides the lack of experience (lack of *how-to knowledge*), the Mozambican energy market suffers from a lack of basic education and understanding of the functionality of a smart grid infrastructure – a lack of so called *principles knowledge,* introduced in chapter four. Respondents from EDM and GIZ assign this lack of principles knowledge to the poor state of formal education in Mozambique.

Staff or experience from mini grid projects in other countries cannot be directly transferred to projects in Mozambique due to the local specificities, as EDM's representatives emphasize. Consequently, additional costs arise for smart mini grid projects since comprehensive capacity building for on-site staff has to be provided.

The lack of economic initiative which was among the revealed barriers in the first part of this study, can also be an important impeding factor for smart mini grids. In a sector which is almost entirely neglected by EDM and only occupied partly by FUNAE, it is essential for smart mini grids that private initiatives, communities or organizations enter the market. Ahlborg and Hammar (2014, 122) observe a relatively low level of community and church initiatives in Mozambique, compared to neighboring countries like Tanzania. Among other cultural and historical specificities, the long-lasting war which lead to many displaced persons and an erosion of social capital might be one of the origins for the relatively low level of community activity in Mozambique. Therefore, to a large extent, the blanks will probably have to be filled by donor-driven or commercial initiatives or by combinations of both. However, commercial activities in Mozambique face the mentioned shortcomings in regulation and governance, as well as the adverse framework conditions such as the poor state of transport infrastructure and import restrictions for the necessary equipment. Import tariffs for mini and off-grid equipment such as solar lamps, converters, batteries or system controllers from non-SADC countries range from 5% to 20% for certain products (DFID 2016, 5). Consequently, a successful implementation of smart mini grids in Mozambique would probably require additional incentives to bolster the diffusion of this innovation.

Due to the several problems, mentioned above, it can be expected that in many cases, smart mini grids in Mozambique will operate only at the edge of profitability. How profitable a project eventually is, depends essentially on regional specificities regarding density of consumers, willingness and ability to pay, commercial initiative and existing know-how (Tenenbaum et al. 2014, 11). As representatives of EDM and Eduardo Mondlane University summarize, smart mini grids in Mozambique will in most cases need subsidies to remain working.

8.3.2. Impact of the drivers

Mini grids and central grid infrastructure are close but not complete substitutes and additional funds can be attracted by mini grid projects which are not accessible for central grid extension. In contrast to central grid infrastructure, mini grids cope much better with the conditions of rural and peri-urban areas. First of all, mini grids do not require large and expensive high-voltage transmission lines like central grids do. Furthermore, the mini grids´ infrastructure is smaller and more flexible in size and capacity. Consequently, a lack of economies of density does not affect mini grids as much as central grid infrastructure. This advantage makes mini grids more feasible for less densely populated areas (Abu-Shark et al. 2006, Blyden, Lee 2006). Therefore, as EnDev´s consultant for solar energy mentions, the relatively low density of population in Mozambique can be a relative advantage of mini grids in comparison to central grid infrastructure.

Addressing the short term orientation of investors, initial installment costs of a mini grid can be kept low by starting with connections of a few households only. Afterwards, the mini grid can be enlarged successively – another advantage of divisible infrastructures (Cescon 2015). Generally, mini grids are a simplification in comparison to a complex central grid system (Platt et al. 2012, 186). Thus, maintenance and administration, especially if supported by intelligent grid management, are facilitated. Data of grid performance is less complex and can be processed more easily (ibid.).

In contrast to mini grids, central grid infrastructure needs an advanced level of already existing economic activity with the corresponding ability to pay to be economically viable. Mini grids can induce this economic activity in areas without any relevant regional added value apart from subsistence agriculture and basic handcraft. Filling some of the blank spots, left by the central grid, mini grids can induce improvements in agricultural, skilled craft and industrial productivity. Thus, electrification by mini grids can lead to more economic development and improve ability to pay for electricity and other commodities in rather remote areas. Due to the strong connections and mutually reinforcing effects between economic development and mini grid implementation, to some extent, power infrastructure can create ability to pay among its customers, as the representatives of the Ministry of Mineral Resources and Energy state. Figure 10 illustrates the synergies between basic electrification and economic development. The process can start at any point of the circle.

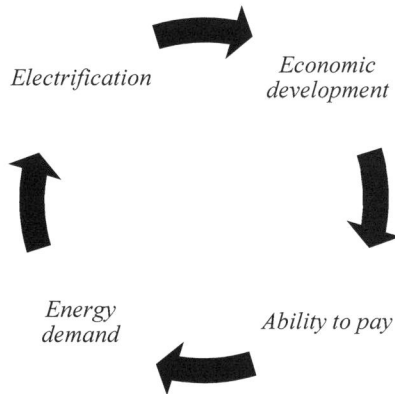

Electrification

Economic development

Energy demand

Ability to pay

Figure 10: *Electrification and rural economic development.*
Source: Own illustration.

Eventually, the synergies between electrification in rural economic development can relieve the lack of ability to pay in rural areas and become a strong driver to new and smarter rural mini grids or even main grid connections.

If the necessary capital is available, rural and peri-urban customers can be expected to have a relatively high willingness to pay for electricity as several respondents state. In unelectrified Mozambican areas, the population usually bears high energy costs using expensive sources such as kerosene, candles, batteries, charcoal or firewood (ministry′s scientific advisor, Chongo Cuamba 2006, 76). In contrast to this, a stable electricity supply is more convenient, healthier and in many cases cheaper (Tenenbaum et al. 2014, 255). This high demand for peri-urban and rural electrification can attract potential investors.

The respondents´ statements indicate that a relatively high willingness to invest in mini grid projects can be expected among FUNAE, private actors and donors. As already stated before, EDM is described by the respondents as reluctant regarding alternatives to central grid electrification. FUNAE, however, can be expected to be more willing to invest in mini grids at the edge of profitability since in contrast to EDM, FUNAE considers itself not as a merely lucrative but rather as a social institution. Its goal is to "give energy also to the poorer families in rural areas", as FUNAE´s head of department claims. A similar case exists for donor-funded mini grid projects: Mini grids are quite likely to benefit from donor-funded subsidies as they meet typical donor goals such as relieving energy poverty with an environmentally friendly and reliable rural electricity supply (Tenenbaum et al. 2014, 120, Sampablo et al. 2017, 35). For example, donor organizations including GIZ offer co-funding for mini grids in Mozambique, GIZ´s representatives explain. In the respondents´ experience, development cooperation can fill initial financing gaps and push new projects. To be sustainable in the long run, though, mini grids need sufficient revenues from their customers, the respondents say.

One factor which positively influences revenue generation for mini grids in Mozambique is the absence of a binding price cap-regulation. Although the regulation of EDM´s tariffs can also influence the price-setting of a mini grid operator as a de-facto regulation (see previous chapter), legally, mini grids in Mozambique possess more flexibility in pricing than EDM. According to EDM´s manager of projects and financing and GIZ´s head of department, the absence of a binding price ceiling acts in favor of mini grid investment. Despite the price-cap´s effect to act as a benchmark for price setting, GIZ´s head of program refers to experience from mini grid projects for which tariffs were defined together with the local community which were higher than EDM´s prices. These experiences indicate that it depends on the specific case, how strongly price-setting for mini grids is influenced by the regulated main grid tariffs.

Looking at the current state of smart mini grid diffusion in Mozambique, it appears that EDM is very reluctant regarding this technology and even FUNAE leaves many potential sites to private investors (Sampablo et al. 2017, 37). Furthermore, private sector engagement is facilitated by possibilities for companies to cooperate with public entities or to act on behalf of the government. Under the public-private partnership law (Law No. 15/2011), generation, distribution and transmission concessions are issued for typically 25 years and power sale concessions for ten years if companies cooperate with local governments and communities (Sampablo et al. 2017, 40). The involvement of local stakeholders – which is required by this law – can improve the perception of ownership and the sustainability of mini grid projects.

As mentioned earlier, digitalization is seen as a strong driver to mini grids in Mozambique. Advanced load and generation control improves the stability of decentral networks which have less possibilities for balancing loads than large grid networks with large generation and distribution capacities (EDM's manager of projects and financing, Ahlborg, Hammar 2014, 123, Tenenbaum et al. 2014, 270). As highlighted by representatives of EnDev, the Eduardo Mondlane University and the Ministry of Mineral Resources and Energy, remote monitoring and diagnosis as well as distant management furthermore allow to operate several mini grids from a central point. Thus, costs can be reduced and effectiveness improved, as EDM's staff emphasizes. Knowing when and how much energy is consumed and generated, generation capacities can be optimized, and storage can be charged or discharged efficiently as GIZ's head of program explains.

Representatives of GIZ and EnDev mention that mini grids can benefit substantially from leapfrogging potentials. Many feasible sites are still not provided with mini grid infrastructure, such that new mini grids can directly benefit from the latest technological developments. The rapid developments have made smart technologies successively easy to handle. Intuitive applications and mobile phone control empower large parts of the population to use technologies which at the beginning of the century still required advanced knowledge of information technology. These simplifications in handling facilitate the introduction of smart mini grids in rural areas with a low level of digital experience. Furthermore, mini grids offer good potentials for re-invention: A small mini grid network can be adapted and modified more easily according to the local needs than large grid systems. Out of all available smart technologies, the ones can be chosen which suit best the local conditions. Thus, compatibility is enhanced.

In Mozambique, some experience with pilot projects for smart mini grids already exists. As FUNAE's head of department mentions, FUNAE already implemented intelligent remote monitoring for some newer mini grids. For these mini grids, production and consumption patterns can be visualized with an interface on a computer screen or a smart phone. One important advantage of remote monitoring is to see if storage units work properly. In mini grids, powered by alternating renewable energies, storage units are essential to deliver constant energy. In effect, FUNAE's newest smart mini grids – a combination of solar panels, batteries, intelligent grid management, wiring and a back-up diesel generator – usually guarantee 24-hour energy supply without staff needed on the site, FUNAE's head of department claims. Representatives of GIZ and Eduardo Mondlane university back these experiences. According to the respondents, pilot projects with remote monitoring and simple automated load management proof successful in Mozambique. According to the mini grid researcher at Eduardo Mondlane University, the implementation of smart meters in a 2 MW mini grid in the province of Niassa showed clear benefits by giving customers and the operator full transparency about consumption, costs and payment profiles.

The expansion of mobile phone use in Mozambique, combined with a good abundance of mobile internet, comes with positive effects on rural electrification by mini grids, too. For example, mobile internet allows for effective machine-to-machine communication in rural areas – a prerequisite for smart grids. Excess capacity of mobile phone towers can be used to transmit information between the grid components and the grid management unit (Smertnik 2015). Furthermore, customers can manage energy use and payments independently using an application on their phone. Thus, billing and information about consumption patterns is improved. Besides increased convenience, mobile billing and consumption control offer cost savings for the mini grid operator, especially in rural areas, as Eduardo Mondlane University's professor of renewable energy mentions.

Another important advantage of smart mini grids is their ability to effectively integrate intermittent renewable energies. Therefore, the vast abundance of primary renewable energy – sun, wind, hydro, biomass – which can be applied at small scale, is undoubtedly a strong driver to smart mini grids in Mozambique, as several respondents say. Szábo et al. (2011, 2) state that the high productivity of renewables – especially solar – in East Africa improves the economic viability of renewably fueled mini grids in this region significantly. It is an important advantage of solar power that it can be deployed basically everywhere in Mozambique as it is not bound to certain spatial or geographical limitations but only sunshine which is widely available all over the country (FUNAE 2014, Sampablo et al. 2017, 11). Experience with solar-powered mini grids exists in Mozambique. For example, the Portuguese and the South Korean governments funded several pilot projects for this technology (Sampablo et al. 2017, 35).

Experience with mini grid installations can also be acquired from Mozambique's neighboring countries. Mini grids are not at all new to Africa. In contrast to Mozambique with its brief history of specific experience in this area, other African countries, such as Tanzania, have already implemented hundreds of mini grids. Mozambique's northern neighbor can be considered as one of Africa's leading countries in mini grid electrification (Sampablo et al. 2017, 47). Due to its comparable socio-economic and geographical conditions, Tanzania can serve as model for mini grid implementation in Mozambique. Accordingly, EDM's and GIZ's representatives underline the large potentials of cross-border networks and the exchange of experience to bolster the diffusion of smart mini grids across the African continent.

The necessary workforce for mini grids in Mozambique can partly be recruited from local communities which live on or close to the sites where the mini grid shall be installed. Usually, basic construction qualifications are available in rural Mozambican villages. Due to a high level of unemployment, labor is easily available. Recruiting local staff has the positive side effect that it can improve the identification of the local population with the mini grid project. Ownership and acceptance of mini grid installations can be further improved if the local communities, local governments or local electrification committees are included into planning and implementation of the projects (Tenenbaum et al. 2014, 79-80). If mini grid operators exclude the local community and exploit their economic power to dictate conditions, severe acceptance problems or even opposition against the project can occur.

Looking at the political impact of decentral mini grids, rural electrification can contribute to mitigate political tensions in Mozambique. Supplying rural areas with electricity can reduce the disparity between rural and urban areas and narrow income gaps. (Ahlborg, Hammar 2014, 123). Especially RENAMO's voter bases which are mainly found in rural Mozambican areas, benefit from rural electrification. Thus, a potential perception of RENAMO-dominated areas to be neglected in the national electrification strategy, can be avoided – an important contribution to national security.

Regarding the performance of the relevant actors, representatives of GIZ and the Ministry of Mineral Resources and Energy, highlight the positive impact of FUNAE´s engagement for rural electrification. Respondents think that it is an advantage that FUNAE holds regional offices. This decentral structure allows FUNAE to be accessible close to the sites of decentral electrification. According to a mini grid market assessment conducted by Sampablo et al. (2017, 11), the enabling environment for mini grids in Mozambique is currently improving due to an increasing awareness of public actors regarding solutions other than central grid extension. Furthermore, the authors claim that the focus of rural electrification has partly shifted from standalone systems to mini grids due to their higher quality of energy supply. This change of mindsets towards mini grids has induced a "substantial progress towards the creation of a policy environment that is supportive of mini grid development", (Sampablo et al. 2017, 11).

8.3.3. Overall feasibility

The analysis above indicates that mini grids in Mozambique are the most feasible option for areas with basic possibilities for consumer bundling but no expectation to be connected to the central grid in the close future. Mini grids cope much better with the conditions in rural areas than main grid infrastructure since mini grids are less affected by missing economies of density and they can be adapted more easily to the specific framework conditions. At the same time, smart mini grids guarantee a stable and high-quality power supply, comparable or only slightly lower than the energy quality of central grid connections.

Accordingly, respondents representing the Eduardo Mondlane University, GIZ and the Ministry of Mineral Resources and Energy explicitly state that mini grids are the best solution for medium populated areas in Mozambique. GIZ´s head of program says that 100% access to the central grid might not even be desirable since decentral solutions such as smart mini grids offer electricity supply which can adapt much better to the specific local conditions in many cases. The respondent expects that still for decades, large regions in Mozambique will remain without access to the central grid. Those gaps can be effectively closed by alternatives to the main grid – such as mini grids – EnDev´s consultant for decentral energy adds.

Recent market assessments for Mozambique support the experts' assessments. Sampablo et al. (2017), supplemented by findings from FU-NAE (2014), find strong potentials for mini grids in Mozambique. The authors estimate that 22% of Mozambique's population, more than 5.5 million people, would be served best by mini grids (Sampablo et al. 2017, 4). The analysis is conducted based on the distance of villages from existing electricity networks and the density of population. High potentials for mini grids are concentrated in the northern provinces, especially Nampula, Zambezia and Tete due to the sparse grid network and sufficient consumer bundling. Further significant potentials exist in the provinces of Cabo Delgado, Inhambane and in rural areas of Maputo province (Sampablo et al. 2017, 13).

The feasibility of mini grids and their quality of power supply has increased a lot due to the advancements in smart grid management in combination with the increasing possibilities to integrate intermittent renewable energies – a power source vastly available in Mozambique. Especially advanced remote monitoring, automated load management and the use of mobile ordering, control and payment are attractive technologies for the Mozambican context. For some of these technologies – such as mobile billing and payment – a substantial level of experience already exists in Mozambique. For the other smart technologies mentioned above, at least first pilot projects have been launched. Additionally, experience can be acquired from neighboring countries and development cooperation as donor organizations show high interest in a smart mini grid electrification.

Despite the structural shortcomings in governance and regulation, Mozambique meets some important institutional requirements for the successful implementation of smart mini grids. First of all, there is FUNAE as a decentrally structured electrification agency with experience in rural electrification with mini grids. Furthermore, mini grid investors face less regulative rigidity than the public utility EDM. Local communities or committees can be included in the implementation process of mini grid projects such that ownership and sustainability can be improved.

Financing remains the bottleneck to electrification in Mozambique, also for mini grids. GIZ's head of program describes financing along with shortcomings in legal security as "killer factors". The profitability of smart mini grids is restricted by the low ability to pay of customers and the public sector. However, financial support, especially from international donors, is available. Subsidies for mini grids are justifiable as energy supply is an essential facility and an important aspect of public development goals. Furthermore, especially renewably powered mini grids are an environmentally friendly alternative to conventional electrification based on fossil sources or large-scale hydropower.

Concludingly, smart mini grids are perceived as relatively advantageous for a significant part of the Mozambican territory, both by the respondents in the interviews and in the existing literature. Additionally, mini grids come with a high level of trialability, observability and a relatively low complexity in comparison to centralized grid infrastructure. Although smart mini grids are a relatively new technology to Mozambique, they offer large potentials for re-invention and adaptation such that they can be made compatible to the economic, political and geographical conditions on the specific sites. The various advantages of smart mini grids for the Mozambican context are reflected by the high interest in this technology among FUNAE, donors and recently also governmental actors including even EDM. Therefore, it can be expected that the diffusion of smart mini grids in Mozambique will continue and presumably even accelerate in the next years.

8.4. Connected smart mini grids

8.4.1. Impact of the barriers

Many aspects of the impact analysis of drivers and barriers for isolated mini grids are also valid for connected mini grids. Therefore, the following reasoning focuses on the specific impacts of drivers and barriers on connected mini grids. This chapter starts with the barriers to the connection of a smart mini grid to the main grid.

To connect a formerly isolated mini grid to the main grid, compatibility of the two systems has to be ensured (Tenenbaum et al. 2014, 13). Several technical prerequisites have to be fulfilled for a connection to be successful. The mini grid´s installations must be able to cope with the high loads and voltages in the main grid. If the mini grids´ infrastructure is not resilient in this sense, serious damage of wiring and technical equipment can occur. On the other side of the connection, the central grid should have sufficient free capacity such that excess energy from the mini grid can be absorbed successfully (Tenenbaum et al. 2014, 156). On the other hand, too much free grid capacity can also be harmful. The scarcity of power in the central grid can be further aggravated by the connection of additional grid capacity unless the mini grid comes with overproportionately large generation capacities. Consequently, the problems of an underpowered main grid can also be transferred to the connected mini grids. Additional grid infrastructure without adequate generation capacity exacerbates the negative ratio of grid capacity and generation capacity. Consequences of the connection of underpowered mini grids to the main grid can eventually be a further increase of blackouts and even more problems in load management. The mini grid operator should consider that connecting a mini grid to an unstable main grid can lead to a deterioration of energy quality for the mini grid customers.

A connection of mini grids changes power flow radically: The unidirectional power distribution network changes to a network in which power can flow into both directions – from the central grid into the mini grid but also from the mini grid back into the main grid. If the main grid has problems in receiving due to an ineffective load management, a lack of capacity or a bad state of infrastructure, the connection of a mini grid creates a challenging situation. If interruptions or even a temporary shutdown in the main grid occur, the connected mini grid´s operator is affected, too, even if the failure is not caused by the mini grid (Tenenbaum et al. 2014, 163).

After the shutdown of the central grid, backup power is needed to get the grid back into service. The restart process takes some time (typically around 15 minutes) to synchronize the power quality – frequency and voltage among others – of the main grid and the connected mini grid (Tenenbaum et al. 2014, 164). To detect failures such that the responsible actor can be sanctioned and to heal interruptions, advanced metering and monitoring is required. The Mozambican power grid is characterized by regular blackouts, transmission losses and a lack of intelligent grid management. The faulty situation in the Mozambican grid network can make mini grid operators reluctant to pursuing the connection with the main grid. It should be considered, too, that the challenge of collecting and processing vast amounts of information becomes even harder if additional interconnections bring additional variables into an already complex and diverse grid system.

Islanding is a possibility to protect a mini grid from instabilities in the main grid. If a mini grid is islanded, it is – usually automatically – disconnected from the main grid, such that the mini grid can be operated like an isolated mini grid. However, islanding comes with several challenges (Platt et al. 2012, 201):

- Grid failures have to be detected in real-time diagnosis or even fore-casted to react rapidly. Thus, islanding comes with high requirements for monitoring, data processing and automated response.
- All controllers must switch immediately from connection mode to dis-connection mode.
- Intelligent load management is needed to match demand and supply in both modes.
- For reconnection, effective re-synchronization is necessary.

Besides these technical aspects, interconnectivity is significantly influ-enced by the regulatory framework. The regulation of connections be-tween the central grid and mini grids is necessary because there is unequal negotiating power between mini grid and main grid operator. The origin of this inequality is the concentration of market power by the national utility EDM in the case of Mozambique. Furthermore, a high level of infor-mation asymmetry can be assumed among the actors. Unregulated bar-gaining power – especially if different levels of information exist – can lead to unfair and economically inefficient terms and conditions.

The quality of regulation determines whether legal and planning securi-ty exist for the case of interconnection. Some important aspects, regulation should define, are the following (Tenenbaum et al. 2014, 156):

- What happens if grid extension takes the central grid to the site of a mini grid?
- Which actor has which financing responsibilities in the case of con-nection?
- Which are the procedures for interconnection?
- Which are the rules for power trading between the operators?
- Which compensations must be granted if one party does not comply to the regulatory requirements?

Respondents criticize that in Mozambique, there is no effective regulation for what happens when central grid extension meets a mini grid. Repre-sentatives of GIZ, Eduardo Mondlane University, EDM, FUNAE and EnDev sharply criticize a lack of reliability and unclear responsibilities in case the central grid arrives at a mini grid´s site. That is, practically none of the prerequisites for an efficient regulation of interconnection is cur-rently fulfilled in Mozambique. Also, the lack of transparency in the plan-ning of grid extensions is a problem since mini grid operators in Mozambique are often surprised by the arrival of the central grid. EnDev´s solar energy consultant shares his experience about such cases: *"With FUNAE, we had [mini grid] projects and on the other side of the road, EDM was coming [with the main grid]. There was not even a minimal level of coordination."*

Also, the regulation of the process of interconnection suffers from significant shortcomings in Mozambique as the respondents´ statements indicate. By law, decree 48/2007 defines the responsibilities and technical requirements for the involved parties which operate a connected mini grid. However, for isolated mini grids, no specific technical regulation – a so called *grid code* – exists (Sampablo et al. 2017, 38, 44) as confirmed by respondents representing the Ministry of Mineral Resources and Energy and the Eduardo Mondlane University. That is, the arrival of the central grid is not anticipated in the regulative framework. If a mini grid is constructed without the expectation to be connected to the main grid one day, the unexpected arrival of the central grid can come with serious compatibility problems or even lead to the abandonment of the mini grid if a connection is not possible. Thus, the lack of technical specifications can lead to an unproductive use of valuable infrastructure and to a waste of resources.

Even if in spite of the several barriers, the connection of a mini grid is successful, involved parties face new regulatory shortcomings after the connection. Most importantly, the question arises which party has which responsibilities. Chapter 7.2.2 described several models for the situation after a mini grid has been connected. If a model is applied which involves an independent operator for the connected mini grid, the mini grid operator and EDM have to conclude on the terms, how power purchase from the mini grid is organized. For this case, respondents representing FUNAE, the Eduardo Mondlane University and the Ministry of Mineral Resources and Energy explicitly demand standardized power purchase agreements which do not exist in Mozambique, so far.

A power purchase agreement is the contract between the mini grid operator and the central utility which defines terms and conditions for how power is traded between the two parties. This also includes quantities of energy to be delivered, payment terms and penalties for under-delivery or other violations of the contract (World Bank 2018 a). If a regulator issues a *standardized* power purchase agreement, the administration of the interconnection is facilitated significantly, regulatory risks are reduced and legal security improved. Since no standardized power purchase agreement exists in Mozambique, the interconnection of a mini grid comes with additional risks and transaction costs for all involved parties. Therefore, as FUNAE's head of department mentions, a working group including EDM, FUNAE, the Ministry of Mineral Resources and Energy and external consultants is currently developing a proposal for a standardized power purchase agreement for connected mini grids. That is, consciousness about the necessity of a standardized power purchase agreement exists. Work is still in progress, though.

Another regulatory barrier is the lack of rules regarding financial incentives for connected mini grids. One effective possibility for an incentive scheme are feed-in tariffs. Feed-in tariffs offer guaranteed prices for electricity fed into the grid by power producers independent from the grid operator. The amount of compensation can be offered independently from the power source or it can be differentiated according to the type of technology, for instance to support renewable energy sources. Feed-in tariffs typically aim to attract additional investors to contribute to the energy supply or to support the deployment of certain technologies (Couture, Gagnon 2010, 955).

Already in 2014, feed-in tariffs were generally approved by the Mozambican council of ministers (decree 58/2014). Initially, the council of ministers planned rates for feed-in tariffs, ranging between USD 0.13-0.41/kWh (Sampablo et al. 2017, 45). A feed-in-tariff in this range exceeds most of the tariff levels, EDM is allowed to apply (see chapter 5.4.2). That is, a feed-in-tariff regime as proposed can change incentives, such that the economic feasibility of investments in connected mini grids increases significantly in relation to main grid investments. However, feed-in-tariffs for the Mozambican energy sector still await implementation (Sampablo et al. 2017, 4), mainly due to governmental budget constraints as respondents explain. According to the representatives of GIZ and Eduardo Mondlane University, the lack of feed-in tariffs clearly inhibits the implementation of mini grids and their connection to the main grid.

Feed-in tariffs can be necessary to compensate for the stricter price regulation in the case of interconnection. As soon as a mini grid becomes part of the main grid, its energy can only be sold under the price cap. If the mini grid operator sells power at retail to its customers, it is directly affected by the price cap (Tenenbaum et al. 2014, 67, Sampablo et al. 2017, 38). Even if the mini grid sells power to EDM and EDM sells the power to the customers, the mini grid operator is indirectly affected since the price cap narrows the final revenues. Thus, EDM's financial power to compensate the mini grid operator for the power fed into the grid is constrained by this regulatory measure. As mini grids often operate at the edge of profitability, influenced strongly by local specificities, the national uniform tariff regime constitutes a challenge for main grid-connected mini grids.

8.4.2. Impact of the drivers

Financial support for the connection of mini grids is generally available from donors, says GIZ´s head of department. FUNAE also offers subsidies for interconnections (FUNAE´s head of department). The more mini grids are connected, the more small and independent power producers feed into the main grid. Consequently, load management becomes more complex. To deal with this additional complexity, automatization of the in- and outflows of loads can help. Accordingly, the ministry´s technical advisor states that the connection of mini grids will make smart control and management necessary.

At the point of interconnection, measuring the amount and quality of energy fed into the grid is a crucial task to manage billing and synchronization of power quality. Especially at this point of interconnection, respondents from FUNAE and EnDev say, smart technologies are required. The respondents highlight the necessity of technologies that ensure sophisticated measurement, remote monitoring and load control.

Several digital innovations support the management at the point of interconnection. One example are digital protective relays. A digital protective relay is a computer-based appliance for interconnections of power grid infrastructure. A software in the relay detects electrical faults and provides metering, communication and self-healing. For example, the relay can automatically break the circuit if a problem (e.g. over voltage) is detected such that damage to the grid is avoided. Additional functions are the monitoring of power quality characteristics such as dynamics, current, voltage or temperature to enable preventative maintenance. Furthermore, the relay controls and processes real time data which can be sent digitally to central control systems, it can receive and execute commands, manage voltage and current, take protective action, send alarms and island a mini grid in case of over or under voltage to protect the equipment (SEL 2017). Nowadays, the application of smart technologies which combine many different functions is cheaper than the traditional way of using separate applications managed by analogue monitoring (Tenenbaum et al. 2014,

217). Joint monitoring of connected mini grids and the main grid further-more reduce the risk of failure or incomplete synchronization such that interconnectivity is enhanced (Tenenbaum et al. 2014, 229).

Since interconnections are facilitated by smart technologies, the high and still increasing coverage with mobile internet in Mozambique is an important driver to push the connection of mini grids. The interconnection of mini grids with each other and with the main grid makes the emergence of a comprehensive smart grid easier than implementing smart technolo-gies on the large scale with a top-down approach. On the lowest layer – the mini grid – smart technologies can be implemented on the small scale with less complexity. If several smart mini grids are connected to each other and to the main grid, they can serve as building blocks for a larger smart grid network which develops step by step from the bottom up.

A mini grid with effective load management integrates all the supply and demand resources within the mini grid such that it becomes a closed system. When it is connected to the main grid, it can be managed as an au-tonomous sub-system of the main grid. In effect, the mini grid can be treated like additional generation capacity which feeds power into the grid but it also absorbs loads if necessary (Platt et al. 2012, 193). The more ef-fectively the potential of smart solutions is used to make the mini grid re-silient and autonomous, the more it can stabilize the overall grid system if connected. A large smart network, consisting of many smart sub-systems exploits the benefits of a large distribution network but also the ad-vantages of a smart grid with decentrally managed loads (Blyden, Lee 2006, 5). In such a system, the failure of one grid component can be com-pensated for by the other sub-systems and robustness against blackouts, reactive power and the curtailment of transmission and distribution losses can be improved (Platt et al. 2012, 187).

In a grid system which consists of a main grid and several connected mini grids, it is even more crucial to identify the responsible actor of fail-ures. Only if errors are detected and sanctioned, all parties have an incen-tive to invest sufficiently in maintenance and reliability. For this end, the ministry's technical advisor urges for the use of advanced monitoring technologies.

The strong presence of donor organizations and consultancies can help to relieve the regulative constraints for connected mini grids. According to GIZ's head of department, a promising area to use the know-how of de-velopment agencies is consulting the Mozambican government in adapting regulation to make it more light-handed for connected mini grids and to close regulatory gaps. As the ministry´s scientific advisor tells, currently, external consultants work together with representatives from the Ministry of Mineral Resources and Energy, EDM, FUNAE and the regulatory agency ARENE to develop a specific regulation for interconnections, not only including standardized power purchase agreements but also a uniform grid code.

8.4.3. Overall feasibility

Especially if combined with smart technology, the connection of mini grids can improve stability, diversify the grid system and improve service quality for the customers. If reliable and fair terms of power trade are concluded on by the mini grid operator and the utility running the main grid, the connection can be beneficial for both sides. Single connections of mini grids can serve as pilot projects such that observability and trialability are relatively high.

However, mini grid connections face serious challenges in Mozambique which make the interconnection costly, risky and difficult. Despite the necessity of regulation, there are insufficient regulative measures to facilitate connections in Mozambique. Precisely, there are no standardized power purchase agreements, no feed-in-tariffs, no uniform grid code and a lack of planning transparency. Put briefly: There are no reliable rules for what happens if the central grid and a mini grid meet. Which will be the responsibilities of the operators? How will power purchase be organized? How does the regulator ensure that one party is compensated for the failures of the other?

Due to the lack of compulsory technical terms, the compatibility of mini grids with the main grid is questionable in Mozambique. The lack of technical specifications also makes comparisons with existing connections difficult or even impossible. Under these circumstances, feasibility is seriously harmed. Many mini grid operators might perceive it as more advantageous to operate rather an isolated than a connected mini grid. Consequently, conditions are not given for Mozambique to be among the early adopters or innovators regarding smartly connected mini grids. It has to be considered, too, that poor chances to connect a mini grid to the main grid can also comprise incentives to invest in a mini grid in the first place. Operators of isolated mini grids might regard it as advantageous if they had the prospect to connect their mini grid to the main grid, one day. Taking all things into consideration, it appears to be significantly more challenging to implement connected than isolated mini grids in Mozambique.

8.5. Smart off-grid solutions

8.5.1. Impact of the barriers

"Most families in rural areas live with less than 30 dollars per month. What of these 30 dollars are they going to use for paying energy?". With this rhetorical question, the ministry´s technical advisor illustrates the low ability to pay of rural customers in Mozambique. Even the relatively low-priced standalone systems – such as the typical off-grid solar home systems – can come with prohibitively high up-front costs for rural households. Hit by poverty traps, these households have little possibilities to save money. Consequently, if standalone systems shall supply large regions with energy, they cannot simply be sold but have to be provided with innovative financing possibilities like payment plans, monthly fees or micro credits (GreenLight Consult 2017, 28). The low ability to pay can be a barrier, especially for profit-oriented actors say the representatives of the Ministry of Mineral Resources and Energy.

The implementation of standalone systems can be further impeded by a lack of abilities to handle and maintain the systems. To illustrate this concern, it might be worth to mention that still broadly half of the Mozambican population is illiterate (UNICEF 2018). Especially in rural areas literacy and basic education are in a poor state. Organizations which pursue a sustainable implementation of standalone systems might therefore be obliged to offer costly training which not seldomly is additionally complicated by language barriers or cultural differences – especially for donors or foreign investors. Initiatives from the private sector in the off-grid market can be further blocked by the regulatory barriers to market entry. The poor state of the regulatory framework also constrains the possibility to interconnect different home systems to a very small mini grid. The representatives of the Ministry of Mineral Resources and Energy especially criticize the lack of basic technical requirements for home systems. In effect, safety and compatibility problems arise.

The severe lack of planning transparency also affects off-grid solutions. If it is unclear if and when a grid arrives at a site, it is difficult to determine whether to invest in home systems or wait for the grid. Home systems can become useless as soon as the grid arrives. On the other hand, home systems can also be combined with a grid connection. For example, if the power supply from the grid is unstable, a home system can be used for back-up such that power supply becomes more reliable for the household. Modern home systems can be connected with the main grid by simply plugging them into a socket. The load management unit in these home systems can synchronize power from the grid with power from the solar panels (SolarWorks 2018 a). Besides improving energy stability, the energy bill can be reduced in this way by replacing some of the grid power with energy from own generation. Nevertheless, planning the implementation of off-grid solutions would be facilitated significantly, if public and private actors made their plans for grid extension and mini grid implementation more transparent.

In general, it should be kept in mind that home systems have a limited capacity. Therefore, the energy quality, home systems offer, cannot be competitive with grid-based power supply as long as the grid works reliably. While stable grid connections supply 24-hour energy which can be used to operate many energy-intensive appliances at the same time, home systems´ storage and generation is usually only sufficient to run a few devices for a limited period of time.

8.5.2. Impact of the drivers

Off-grid solutions are built to supply one household or one small business. Consequently, they do not require any grid infrastructure and they can be adapted easily in capacity such that each system fits the specific consumer demand. Thus, standalone systems cope very well with the lack of economies of density in rural areas. In contrast to grid-based infrastructure, characterized by large indivisibilities, off-grid solutions can be regarded as very feasible for remote areas with a low density of population.

Like mini grids (see chapter 8.3.2), standalone systems can create additional ability and willingness to pay if the additional energy quality leads to more productivity. Even though off-grid systems do not offer the same level of energy supply like grid-based systems, they can improve production especially in rural Mozambican areas, nowadays still dominated by very basic forms of subsistence farming. For example, the new and better power supply can be used to improve irrigation or refining of raw products, as the ministry´s technical advisor mentions.

Projects for off-grid electrification are very likely to receive funding from donor organizations because they meet typical donor goals such as low-impact extension of electricity access, dissemination of renewable energies, rural electrification and the relief of energy poverty (Tenebaum et al. 2014, 120). For example, the British development agency DFID implemented the USD 45 million program "BRILHO" in Mozambique which intends to expand market access for off-grid technologies along with promoting mobile pay-as-you-go systems (DFID 2016, 4). Additionally, renewable off-grid projects are viable for funding from support schemes concluded on under the Kyoto Protocol (UN 1997) and the Paris Agreement on climate change (UN 2016) to promote climate change mitigation. According to GIZ´s representatives, development cooperation can fill initial gaps and push new projects for off-grid electrification. The respondents qualify, though, that to be sustainable, there needs to be sufficient willingness and ability to pay among the users, such that operation and maintenance can be financed in the long run.

Besides donors, FUNAE offers subsidies for off-grid electrification. According to FUNAE´s head of department, the organization pursues social goals and aims to guarantee a basic level of energy access for poor people in rural areas. Therefore, FUNAE engages actively in distributing solar home systems, the respondent says. The off-grid activities run by FUNAE and donor organizations in Mozambique generate experiences and learning effects. If such experiences prove positive, the development in the off-grid sector can pull additional actors from the private sector into the Mozambican energy market, EnDev´s solar energy consultant says.

For off-grid systems, power tariffs are usually not relevant. Typically, standalone systems are not paid for by a price based on consumption. Instead, the entire system itself is bought or leased. Therefore, off-grid systems circumvent the barriers arising from low and price-capped tariffs in Mozambique. Consequently, to evaluate whether there is sufficient ability and willingness to pay for off-grid systems, one has to take a look at the total costs of a standalone system. In Mozambique, the willingness to pay for off-grid systems can be assumed to be relatively high. As consumers, the adopters directly benefit from the clean, convenient and relatively stable source of energy (GreenLight Consult 2017).

Ability and willingness to pay for solar home systems were analyzed by GreenLight Consult (2017) in the district of Moamba, Maputo province – a typical rural district with a large proportion of households without connection to the electricity grid. The study analyses the implementation of solar home systems with one-time deposit costs of 1750 meticais (USD 23) and a daily fee of 25 meticais (USD 0,33) paid for one year. That is, total costs of the system excluding the deposit equal 9.125 meticais or broadly 125 US dollars. These basic home systems consist of a combination of wiring, a solar panel, a storage and management unit, lightbulbs, torches, a phone charger, a radio and further sockets.

The study shows, that the vast majority of the households is able and willing to pay the deposit which is regarded as necessary to maintain payment discipline. In a poll among 80 randomly chosen households from Moamba district, 67% said that they can and want to pay the deposit, 13% ticked "maybe" and for 20% the deposit is too high (GreenLight Consult 2017, 26).

Regarding ability to pay for the monthly fee, the study compares current household expenditures for energy with the variable costs of solar home systems. According to the GreenLight analysis, without access to electricity, expenses for energy services which could as well be supplied by home systems, are estimated for an average household at 689 meticais per month. For example, people spend money on kerosene for lighting, batteries for the radio or shops where they can charge their phones. With a daily fee of 25 meticais, monthly payments for the home system equal approximately 750 meticais which is only slightly higher than energy expenses without the home system. That is, even if one looks only at the bare monthly expenditures, home systems are competitive. Hence, the average household in the study can be expected to be able to pay for these basic home systems (see also: DFID 2016, 1).

Additionally, the home system comes with significant additional benefits such as time savings and more convenience. For example, in rural Mozambican areas, a large share of the population pays to get their phone charged. The following example from Moamba district illustrates the corresponding costs and inconveniences (Greenlight Consult 2017, 19 f.): 37% of the 80 participants in the poll say that they pay others to charge their phone due to the absence of own sources of electricity. These phone users spend on average 40 minutes to travel back and forth just to charge the phone. In many cases, travelling also produces direct additional costs, for example for public transport. Only 45% of the respondents charge at home, 38% charge at another person´s home, 13% in the local market and 4% at a private vendor. Rates for charging the phones typically range between one and 10 meticais (USD 0,01 to 0,13) and most commonly, respondents charge their phone every three days. Considering the high penetration of mobile phones in Mozambique and their high importance in peoples´ lives, already this single example illustrates the significant charging and travelling costs which can be saved by the application of a home system.

While grid extension and mini grids typically reach regions where some kind of electricity supply has already existed before (e.g. home systems, car batteries, generators), new home systems are typically supplied in areas which nearly exclusively use traditional fuels such as firewood or charcoal. Some benefits of home systems in contrast to traditional fuels are (DFID 2016, 7 f.):

- Education (extending studying hours through good and stable lighting, access to information sources such as radio, TV, internet)
- Health (less pollution)
- Quality of life (more time for family, leisure, entertainment programs)
- Deforestation is avoided

The additional benefits described above can be assumed to be associated with additional willingness to pay in comparison to traditional energy sources such as firewood, charcoal, kerosene or car batteries (GreenLight Consult 2017, 29).

By the time, the sum of the paid daily fees reaches the total costs of the system, it belongs completely to the household. From this point, the household does not have to bear any expenditures for the related energy services any more if no severe maintenance expenditures arise. This can be an additional motivation to purchase a solar home system. Accordingly, in the poll in Moamba district, 80% of the respondents said that they can and will pay the 25 meticais daily fee, 13% ticked "maybe", 7% regarded the fee as too expensive (GreenLight 2017, 26).

Especially in rather poor rural areas, such as Moamba district, ensuring willingness to pay depends crucially on the possibility to pay for the home system in monthly fees. Ability and willingness to pay can be expected to be insufficient in most cases if total system costs have to be paid at once. Progress in digital technologies makes payments in daily or monthly fees possible, even across long distances. Therefore, the viability of off-grid systems depends crucially on the availability of mobile pay-as-you-go technologies. Here, the user pays the daily or monthly fee via mobile payment to keep the system running. If the user cannot pay or does not need the system, he or she just stops paying and the system is automatically switched off. As soon as the user gets back to paying, the system is automatically switched on again. Consequently, the user only pays if he or she wants to use electricity and if he or she can pay for it.

Pay-as-you go technologies facilitate paying substantially, reduce transaction costs and create effective incentives to make the required payments as representatives of GIZ, EnDev and the Ministry of Mineral Resources and Energy emphasize. The scientific advisor at the ministry of energy concludes that these improvements in consumer handling are strong drivers to the diffusion of pay-as-you-go off-grid systems in Mozambique.

It becomes evident that smart technologies such as mobile payment, remote control and pay-as-you-go are core enabling factors for the dissemination of off-grid solutions in Mozambique. If these technologies were not available, high costs for payment collection, travelling, staff and monitoring would be induced. Although, the use of mobile payment technologies such as M-Pesa, M-Kesh or E-Mola is not as high in Mozambique as in other African countries, the study by GreenLight Consult (2017) shows that there is a considerable level of openness to these technologies. In the poll conducted in the district of Moamba, 91% of the respondents said, they would be willing to use mobile payment to pay for their home system (GreenLight Consult 2017, 27).

Smart technologies also facilitate the operation of off-grid solutions for the supplier. Remote monitoring shows which systems work and which ones require service. Thus, on-site staff can be contacted immediately if problems occur and a high level of functionality and quality can be maintained. Especially in a large country with poor infrastructure like Mozambique, remote monitoring facilitates the operation of off-grid systems, says GIZ´s head of program.

Modern home systems come with a basic level of automatic load management within the system. A control unit autonomously charges the storage unit if surplus energy is produced by the generation unit – typically the solar panel – and discharges the storage if more power is consumed than produced (SolarWorks 2018 a). Thus, reliability of power supply is improved and the viability of intermittent renewable energy sources for home systems is enhanced. Furthermore, smart home systems protect themselves from overloads or deep discharge and manage charging and discharging intelligently such that battery life is extended (Victron Energy 2018). Since many regions in Mozambique are still unsupplied, they can directly benefit from the latest advancements in technology and design. That is, leapfrogging opportunities are also relevant for off-grid systems.

Looking at the economic environment for off-grid electrification in Mozambique, there are some drivers which can bolster the diffusion of this technology. For the majority of rural Mozambican villages, it can be expected, that there will be no grid-based power supply, even in the long run. The economic framework conditions such as missing consumer bundling and a lack of funds constitute significant barriers to mini grids and especially to main grid electrification. For instance, only 1% of respondents in the poll conducted by GreenLight Consult in Moamba (2017,12), believed that the main grid will arrive at their village in the next five years, 66% answered, they had no idea when the grid would arrive and 30% believed that the grid will never arrive. Consequently, there is much free space for off-grid systems in Mozambique's rural areas. If the Mozambican government wants to achieve its goals for electrification and include the rural population, at the current state of technological development, it has no viable alternatives but to support the implementation of off-grid systems (DFID 2016, 1).

Some of the equipment for standalone systems can be acquired from local Mozambican companies. For example, the company Solarkom which is partly operated by FUNAE, produces solar panels and assembles solar home systems in a factory in Matola close to Maputo. Besides, for decentral installations, there is sufficient workforce available in the villages which are supplied. Since home systems are less complex than grid-based infrastructure, the installation of home systems does not require specifically qualified staff, such that members of the local community can be employed for installation and basic maintenance after some basic introduction into the technology. New installations can benefit from rich experience as off-grid systems have already been implemented in various areas in Mozambique, especially driven by donors and FUNAE (Sampablo et al. 2017, 35).

Home systems, which are typically fueled by solar power, benefit a lot from the vast abundance of solar radiation which has already been extensively described in the previous chapters. Solar-based electrification is more resilient to climate change than alternative approaches such as mini hydro generation and more sustainable than fossil fuels. Therefore, EDM's manager of projects and financing states, solar home systems in Mozambique will become even more attractive in the next years.

In contrast to grid-based infrastructure, home systems can usually be applied on houses built from materials like sticks, mud, reeds, bamboo or palmleaves. That is, home systems are compatible with traditional

Mozambican building techniques and do not require costly upgrades of the buildings which grid-based infrastructure sometimes does.

Concerns regarding the lack of legal security do not affect off-grid systems as much as grid-based infrastructure. Investment costs are not that high and amortization does not take that long for off-grid solutions. For a short period of time, regulatory and legal risks can be assessed more easily than for long periods of amortization. Additionally, off-grid systems are not in the focus of regulation which leaves more economic freedom for investors as GIZ´s head of program states. While conventional power grids are regional natural monopolies which require effective regulatory interventions by the government, off-grid systems are a very decentral and small-scale infrastructure which does not automatically induce a relevant level of market power if different suppliers are available.

It should be kept in mind that rural electrification can only be successful and sustainable if the local community is included such that no acceptance problems arise (Tenenbaum et al. 2014, 79). Off-grid systems benefit from the generally high acceptance of electricity infrastructure in Mozambique, as first experiences show (GreenLight Consult 2017, DFID 2016).

8.5.3. Overall feasibility

As the preceding analysis and a high level of expert agreement show, off-grid home systems are the best and the only feasible option for sparsely populated and relatively poor areas in Mozambique. According to a study conducted by DFID (2016, ii), in Mozambique, the average costs per connection for off-grid electrification are only about 6% of the average costs of a connection for grid-based electrification. Eduardo Mondlane University´s professor of renewable energy concludes *"In these regions [areas with very dispersed population], there is no other option but isolated systems."* GIZ´s head of program specifies: *"The lower the density of population, the more off-grid solutions make sense."*

In line with these assessments, studies (DFID 2016, GreenLight Consult 2016, Sampablo et al. 2017) and respondents in the interviews state that large areas in Mozambique offer good conditions for off-grid solutions to be the fittest of all available solutions. The ministry´s scientific advisor reminds that about two thirds of Mozambicans live in rural areas while a large part of these areas does not offer conditions for grid-based infrastructure. Accordingly, respondents expect many areas in Mozambique to remain unelectrified by grid infrastructure such that "this gap will have to be closed by decentral solutions" as EnDev´s consultant for decentral energy says. The inference that standalone off-grid systems are the most viable option for sparsely populated areas is supported by findings from Sampablo et al. (2017, 4). The authors show that for at least 31% of the unelectrified Mozambican population, off-grid systems would be the most viable option.

One important reason for this clear relative advantage of off-grid solutions for poor rural areas is their relatively low price which can be expected to be affordable by a large share of rural Mozambican households. The advantages of home systems compared to traditional fuels and the additional benefits of a convenient electricity supply are expected to induce a high willingness to pay among unelectrified households. Additionally, the contribution of off-grid systems to the relief of energy poverty and to an environmentally friendly power supply unlocks funds from donors and public agencies. Furthermore, the implementation of off-grid solutions circumvents most of the regulatory barriers, political tensions and governance problems in Mozambique. Off-grid solutions cope well with the conditions in rural areas – a lack of consumer bundling and traditional building techniques – and are relatively easy to handle. The solar energy consultant, therefore, describes off-grid systems as a "quick and easy way for electrification".

Off-grid solutions have experienced vast improvements in feasibility due to the advancements in information technology such as automated load management, remote control and digital payment – an electrification with modern home systems is necessarily a smart electrification.

Due to their small size, off-grid systems can be tested in small scale projects such that experiences can be used for improvements in future projects. By the time one household applies a solar home system, neighbors and other villagers will see the system and probably be interested in their services and advantages. That is, trialability and observability are high for off-grid systems.

Taking all things into consideration, for poor rural areas, off-grid solutions come with high levels of relative advantage, observability, trialability and compatibility. Additionally, off-grid systems are typically less complex than conventional grid-based solutions. A lower level of complexity facilitates handling and maintenance. Consequently, prerequisites for a relatively quick diffusion of off-grid systems can be expected to be given in Mozambique. However, to further facilitate the diffusion of off-grid solutions in Mozambique, EnDev´s representatives demand additional studies which analyze the feasibility of off-grid solutions at specific sites along with the systems´ long-term impact on well-being, productivity and education.

As EnDev´s representatives conclude, off-grid solutions are a very suitable innovation for the first phase of electrification – which is the current state of electricity supply in many Mozambican regions. In an assessment of off-grid solutions´ feasibility, it should not be neglected, though, that off-grid systems do not offer the same energy quality as grid-based systems. Off-grid systems will probably get replaced by grid infrastructure as soon as ability to pay and density of population increase. Consequently, it can be assumed that there is a bell-shaped correlation between economic development and the application of additional home systems. The first phase of economic development associated with additional ability to pay makes home systems affordable such that traditional energy use will probably be replaced by electricity from home systems. If economic development continues such that ability to pay and energy demand further increase, it can be expected that home systems are replaced by grid-based connections which offer a more convenient power supply. The assumed bell-shaped correlation between economic development and the additional installations of home systems distinguishes off-grid solutions from grid-based connections, since grid-based connections tend to continue increasing if economic development proceeds until the market is saturated.

Bringing electrical energy to areas with no access to modern energy whatsoever, off-grid systems can contribute to more social inclusion. Access to electricity and modern communication technologies is necessary for social participation. If only densely populated areas with sufficient willingness to pay were electrified, the resulting inequality can easily lead to the perception of injustice and political tensions, FUNAE´s head of department says. Concludingly, home systems support the harmonization of living conditions between urban and rural areas in Mozambique.

8.6. Concluding remarks: Which role for which option?

In the preceding chapters, the impact of drivers and barriers was analyzed for each option for a smart electrification. Table 6 and table 7 summarize these findings and compare the impacts of drivers and barriers with each other. Thus, it can be clearly seen how each option copes with the conditions in Mozambique in comparison to its alternatives.

Table 6 shows whether an option is barely affected (++), moderately affected (+), strongly affected (-) or very strongly affected (--) by a barrier, based on the results of the preceding chapters. The table indicates that the different barriers to electrification in Mozambique affect especially a smart central grid which impedes grid extension and the implementation of smart technologies in the main grid system.

Table 6: *Impact of barriers.*

Barrier	Central smart grid	Smart mini grids	Connected smart mini grids	Smart off-grid solutions
Finance				
Lack of capital in general	--	-	-	+
Lack of domestic capital	--	-	-	+
Low willingness to pay	-	-	-	++
Short-term orientation	-	+	+	++
Electricity market				
High level of market power	-	-	-	+
Low revenues	--	-	-	++
Low power tariffs	--	-	-	++
Lack of tariff flexibility	--	-	-	++
Transaction costs	--	-	-	+
Economic environment				
Poor infrastructure	--	+	+	++
Lack of qualified staff	-	+	+	+
Low density of population	--	+	+	++
Lack of initiative	-	--	--	--
Environmental impact	--	+	+	++
Infrastructure				
Indivisibilities	--	+	+	++
Lack of equipment and tools	--	+	+	+
Lack of maintenance	--	-	-	-
Missing real-time information	--	+	+	+
	-	+	+	+
Economic performance				
Higher costs of smart grid	--	+	+	+
Governance				
Political tensions/conflicts	--	-	-	+
Performance of institutions	--	--	--	+
Centralized decision-making	--	++	++	++
Corruption	--	-	-	-
Regulatory framework	-	-	--	+
Unclear responsibilities	--	-	--	+
Lack of ICT-knowledge	--	-	-	+
Lack of energy politics knowledge	--	-	--	+

Source: Own table.

Smart mini grids are also affected, especially by a lack of funds, low revenues and shortcomings in the political-economic environment. However, mini grids cope much better with the conditions in less densely populated rural and peri-urban areas than main grid infrastructure. Furthermore, their smaller scale exploits the benefits of a less centralized decision making and the shorter amortization time makes it easier to deal with short-term orientation.

Although isolated and connected mini grids only differ in the impact of a very few barriers, these presumably slight differences have large and serious impacts. Interconnected smart mini grids are hit strongly by regulatory shortcomings and the absence of clear responsibilities. These barriers represent very problematic characteristics of the Mozambican energy sector like a lack of legal security, a lack of planning security and a missing quality regulation of infrastructure. Consequently, as the preceding chapter showed, the relative advantage of connected mini grids suffers substantially in relation to isolated mini grids.

Smart off-grid solutions are hit least by the barriers because they circumvent most of them. Most importantly, off-grid solutions are per definition not affected by the problems of grid-based infrastructure, such as indivisibilities, price-cap regulation and market power among others. Furthermore, their small scale, decentral character and their lower level of complexity make them less vulnerable to regulatory shortcomings, governance problems or adverse conditions in the economic environment. Additionally, their relatively low price in combination with intelligent payment schemes (e.g. pay-as-you-go) makes them affordable for a large share of the population and relieves the problem of short capital.

Table 7 shows if an option benefits very strongly (++), strongly (+), moderately (-) or barely (--) from a driver. The table indicates that all options benefit from the drivers. Especially mini grids can use all the available drivers in the areas of finance, economic environment, potential of digitalization, economic performance, governance and acceptance.

Since expert analysis indicates that most donor organizations rather favor decentral options, central grid extension is expected to receive less support from development aid in relation to its alternatives. Furthermore, as a natural monopoly, the central grid is not expected to benefit from the increasing private sector engagement. Especially off-grid solutions but also mini grids benefit from the digitalization in the energy sector and from the abundance of mobile internet which makes the use of these digital innovations possible, even in remote areas. The various potentials of digitalization to improve the feasibility of off-grid solutions reduces the competitivity of the central grid.

Table 7: *Impact of drivers.*

Driver	Central smart grid	Smart mini grids	Connected smart mini grids	Smart off-grid solutions
Finance				
Availability of foreign capital in general	+	++	++	++
Availability of donor capital	+	++	++	++
High willingness to pay among donors	-	++	++	++
Economic environment				
Economic growth	++	+	+	-
Industrialization	++	+	+	-
Increasing agricultural production	+	++	++	-
Growth of energy demand	++	++	++	+
Increasing private sector engagement	-	++	+	++
Abundance of mobile internet	-	+	+	++
Potential for renewables	+	++	++	++
Connection of public facilities	+	++	++	-
Potential of digitalization				
Potential to reduce transaction costs	++	++	++	++
Potential to improve maintenance	+	++	++	++
Increasing use of ICT	+	++	++	++
Already existing smart energy projects	+	++	++	++
Potential to improve decentral solutions	--	++	++	++
Improve energy supply in general	+	++	++	++
Leapfrogging due to new technologies	+	++	++	++
Economic performance				
High benefit of electrification in general	++	++	++	++
High benefits of smart solutions	+	++	++	++
Potential of cost reduction for utilities	+	+	++	-
Leapfrogging due to cost effectiveness	++	++	++	++
Reduction of transmission losses	+	++	++	--
Governance				
Government´s electrification goals	++	+	+	+
Development cooperation	+	++	++	++
Acceptance				
of infrastructure projects	++	++	++	++
of smart energy solutions	+	++	++	++

Source: Own table.

On the other hand, off-grid solutions per definition do not benefit from the potentials of smart technologies for grid infrastructure, such as the reduction of transmission losses or cost reductions for grid operators. Furthermore, off-grid solutions can be expected to be replaced if economic development leads to higher incomes in rural areas because demand for high-quality grid-based connections can be expected to increase with the ability to pay. Therefore, the competitivity of off-grid solutions can be expected to decrease if industrialization and increasing agricultural production lead to economic growth that also benefits poor rural areas.

What follows from the different options´ characteristics and from the way, they deal with the drivers and barriers, are different performances regarding the core factors for innovation diffusion: Perceived relative advantage, compatibility, trialability, complexity and observability (see chapter four). The better an option comes off regarding these factors, the higher its rate of adoption in Mozambique will probably be.

Tables 8 and 9 present a brief overview, summarized from the analysis of the overall feasibility of each option in the preceding chapter. These tables and their underlying analysis allow for two inferences:

1. Smart grid and smart off grid options are feasible in Mozambique and they fulfill essential conditions for a successful diffusion in the Mozambican energy sector.
2. The feasibility of each option depends strongly on the local conditions, especially density of population, proximity to the main grid, ability and willingness to pay as well as availability of external funding.

That is, all options for smart electrification are relevant for Mozambique. Accordingly, FUNAE´s head of department advocates a mixed strategy with a focus on central grid supply in urban areas with a high energy demand and mini grids for rural areas with supplementary solar home systems for areas with a very dispersed population. "The combination is the right way.", is how EDM´s manager for projects and financing puts this analysis into a nutshell. All the other experts and actors, consulted on this issue, also agree with the assessment that a simultaneous implementation of the different options with a regional differentiation is the best strategy for a smart electrification in Mozambique.

The question arises which share each of the options can be expected to have in Mozambique´s future energy supply. Results from a study by

Sambablo et al (2017) indicate that 22 per cent of the Mozambican population will best be served by mini grids and again around 22 per cent by standalone systems (Sampablo et al. 2017, 4). Due to the various barriers and challenges, it is questionable, though, if this large potential of decentral solutions will be exploited effectively.

Table 8: Comparison of options. Part I.

	Perceived relative advantage	*Compatibility*	*Trialability*	*Complexity*	*Observability*
Central smart grid	High in densely populated areas with high ability to pay Low in rural areas Impeded by organizational, regulatory and governance-related shortcomings	Compatible with existing experiences Smart technologies rather applicable within new infrastructure (leapfrogging)	Low due to indivisibilities and high specificity of infrastructure	High: Very complex system with many different and interconnected components	High but this does not push diffusion since no other actors but EDM can enter the market (natural monopoly)
Smart mini grids	High in peri-urban and rural areas with basic level of consumer bundling Implementation facilitated by less tariff rigidity and support from donors	Fairly compatible. Potential for re-invention and possibilities to adapt to local conditions. However, relatively new technology for Mozambique.	High, due to smaller scale than central grid	Medium level of complexity.	High, since mini grids consist of large visible infrastructure in the public space. Thus, they can attract additional adopters.

Source: Own table.

Table 9: Comparison of options. Part II.

	Perceived relative advantage	Compatibility	Trialability	Complexity	Observability
Connected smart mini grids	Relatively low. Despite large potential benefits. Interconnection is especially impeded by a lack of legal security, missing planning transparency and a poor regulatory framework.	Restricted compatibility Relatively new technology for Mozambique Lack of grid code comes with risks for compatibility	High, due to smaller scale than central grid	More complex than isolated mini grids but still less complex than main grid infrastructure	Relatively high Like mini grids, infrastructure is visible. Potential adopters can hardly distinguish between isolated and connected mini grids, though.
Off-grid solutions	High for rather poor and sparsely populated areas	Compatible with traditional building techniques but a relatively new technology First experiences already exist	Very high due to small scale and low-cost implementation	Lowest level of complexity	High Components are visible in public space and can be expected to raise interest among neighbors and further villagers.

Source: Own table.

FUNAE´s head of department expects that in 20 years, 60 to 70 per cent of the total electricity access will still be through the central grid. Accordingly, the respondent expects decentral solutions to contribute around 30 to 40 per cent of the connections. GIZ´s head of program agrees that in the closer future, "a large share of the population" will rely on small scale decentral solutions. The respondent qualifies that solar home systems, nowadays, are only in the beginning phase of adoption. *"Someday, there will be a large push"*, the respondent says. Speaking in terms of diffusion theory, the GIZ representative expects that home systems have not yet reached the take-off phase (see chapter 4.1). The respondent reminds, though, that due to the challenging environment, also for off-grid options, Mozambique is not a "low hanging fruit".

The analysis in the preceding chapters showed, too, that especially mini grids and off-grid solutions offer large potentials for digitalization or even depend on smart solutions to be viable. Consequently, the large expected share of decentral options in Mozambique´s future power supply supports the assessment, that Mozambique´s power sector is already experiencing and will further experience a substantial smart upgrade.

Additionally, also central grid electrification has already experienced first steps of becoming a smarter grid. This development can be expected to continue. If the conditions for the connection of mini grids become more attractive in the future, increasing interconnections will make further smart upgrades necessary. The connection of mini grids or the interconnection of former off-grid solutions can lead to the inclusion of more and more smart components into the overall distribution system. Thus, a smart upgrade of the Mozambican energy sector can grow from the bottom up, even if political decision-makers neglect the potentials of smart technologies.

That is, most likely, the question is not if Mozambique´s future energy sector will be smart but to what extent it will be smart. In the words of the respondents in the interviews, EDM´s manager of projects and financing expresses that making Mozambique´s power sector smart is a desirable "technical revolution" and EnDev´s consultant for decentral energy states that "digitalization is the way" for all options of electrification.

While some smart technologies are expected to be well received in Mozambique – remote monitoring and control, automated load management, remote billing among others – the implementation of consumer participation remains difficult. In this context, the renewables scientist demands not to push consumer participations artificially with measures like compulsory smart meter rollouts. Instead, the respondent advocates a demand-driven implementation. If economic framework conditions change, the respondent says, the demand for advanced metering technologies will automatically increase.

What are the central findings to memorize from this chapter?

- Despite the challenging conditions in the Mozambican power market, it can be concluded that Mozambique will most likely experience the continuation of a parallel implementation of different options for electrification.
- A special and increasing focus can be expected to rest on decentral options such as mini grids and off-grid solutions.
- Findings indicate that electrification in Mozambique will be accompanied by an increasing adoption of smart technologies.

For decentral smart solutions, the pace of diffusion can be expected to further increase in the next years. Especially organizations engaged in the off-grid sector benefit from a demand-driven acceleration of home systems adoption.

9. Policy recommendations

In the second field research, each of the respondents used the interview to express certain demands addressed to the political decision makers in Mozambique. Policy recommendations especially look at the following six areas:

- Performance of public institutions,
- Quality of regulation and legal security,
- Access to information,
- Support for digitalization,
- Quality of maintenance,
- Education and qualification.

Table 10 summarizes the policy recommendations mentioned by the respondents. How often a recommendation was mentioned should not be narrowly interpreted as the importance of a recommendation. Some recommendations responded directly to questions from the interviews while others were brought up independently from the interview guide.

Regarding the performance of public institutions, respondents especially demand a better coordination between the relevant actors, a higher effectiveness of the public programs to push electrification and more reliability of public institutions. Respondents engaged in decentral electrification (FUNAE and EnDev) also demand a change of mindsets regarding decentral and renewable energies. While these statements might certainly be influenced by personal or institutional interests, they can nevertheless be interpreted as an expression of discontent with the electrification strategy of the governmental institutions which still focus strongly on conventional power supply and main grid extension. It is furthermore worth to emphasize that even a representative of the Ministry of Mineral Resources and Energy expresses the need for a better coordination between the public actors in the energy sector. That is, to some extent, awareness of the problem of missing coordination has already reached the relevant actors.

Table 10: *Policy recommendations.*

Identified recommendations extracted from expert interviews. 1 = EDM, 2 = FUNAE, 3 = Ministry of Mineral Resources and Energy, 4 = EnDev, 5 = Donor, 6 = Scientist renewable energy, 7=Scientist mini grids. Number of interviews with each organization: $i - ii$. The Total $\sum$ refers to the number of interviews in which the recommendation was mentioned by the respondent.

Policy recommendation	1	2	3	4	5	6	7	$\sum$
	i	i	ii	ii	i	i	i	
Improve regulation	X	X		XX	X	X	X	7
Intensify implementation of digital solutions	X		X	X	X	X		5
Standardize power purchase agreements			X	XX		X	X	5
Improve legal security	X	X			X	X		4
Conduct regional viability studies	X	X				X	X	4
Change of mind regarding renewables			X	XX				3
Create attractive conditions for investors	X					X	X	3
Improve coordination			X	XX				3
Improve transparency of grid planning	X				X	X		3
Provide sufficient subsidy schemes	X	X		X				3
Get external consulting	X		X					2
Introduce flexible tariffs				X	X			2
Improve access to reliable energy sector data		X		X				2
Improve effectiveness of public actors		X		X				2
Improve reliability of governance		X				X		2
Remove import restrictions for equipment	X			X				2
Set clear rules for interconnection	X	X						2
Close remaining gaps in mobile network		X						1
Focus on projects with synergies			X					1
Improve basic education	X							1
Improve maintenance	X							1
Extent the main grid´s distribution networks				X				1
Specific Power Purchase Agreements						X		1
Teach more specific energy sector issues	X							1
Use local expertise	X							1

Source: Own table.

According to the renewables scientist, the public actors in Mozambique should especially focus on difficult sites which require subsidies and leave the more feasible sites to private actors. Thus, the respondent says, electrification can be accelerated: Private actors electrify the more feasible sites while the governmental actors can focus their resources on sites which would most probably not be electrified by private actors. Actors from the public sector might be reluctant to this recommendation. Leaving the most feasible sites to private actors would mean that the public institutions – that is the tax payers – bear the losses while most of the benefits are generated by private actors. On the other hand, from an ordoliberal perspective, it could be argued that public actors should only engage in operative economic activity if the market fails to produce the desired commodities, such that efficiency and total wealth is increased (Eucken 1965, 151-195).

Regarding grid extension, EnDev´s decentral energy consultant recommends that EDM should not be satisfied with extending the central high-voltage grid to a certain district. Instead, the respondent motivates EDM to install also the necessary regional distribution network quickly after the extension of the large transport lines, such that all households benefit from the arrival of the grid. The ministry´s technical advisor encourages the relevant actors to identify and support electricity projects which can create synergies with economic development to improve well-being and ability to pay. As an example, the respondent proposes to supplement electrification with projects for improved irrigation and agricultural production such that the benefitted households can generate an additional surplus. The necessary condition is that households know how the new energy source can be used productively. Here, governmental experts but also donor agencies or consultancies can provide for the necessary transfer of skills.

To ensure a sustainable impact of electrification projects, the quality of infrastructure has to be kept at a high level. However, the analysis of drivers and barriers revealed some shortcomings in operation and maintenance. Correspondingly, EDM´s staff states self-critically, that maintenance should improve. To achieve this goal, different strategies or combinations of them are available: Channeling more resources into maintenance, improving the qualification of the relevant staff or improving coordination and efficiency of the organization of maintenance. This can include internal reforms among the relevant actors.

Representatives of EnDev and the Ministry of Mineral Resources and Energy encourage to rethink the institutional structure in the Mozambican energy sector. According to the respondents, competences and responsibilities should be separated such that there is one "real regulator", as EnDev's solar energy consultant puts it. Nowadays, both experts criticize, responsibilities are unclear, especially regarding regulation. EDM's manager of projects and financing calls CNELEC which is supposed to become the central regulator under the new name ARENE (see chapter 2) "incapable". The respondent welcomes that it was decided to create a specialized regulatory agency but criticizes its lack of competences, abilities and experience. A transfer of knowledge and competences to the new regulator is needed to make it strong and effective, the respondent says.

Besides reforming the institutional framework, respondents urge for substantial changes in the operative regulatory system. "Improve regulation" was mentioned by seven of the nine respondents. Only the representatives of the Ministry of Mineral Resources and Energy – who would have criticized their own authority – did not express this demand explicitly. Nevertheless, even the ministry's staff contributed proposals for specific changes in regulation.

Especially representatives of organizations affected by regulation proposed several precise approaches for improvements. Two of the most prominent demands are improving the state of legal security and the standardization of power purchase agreements – a demand, connected to "set clear rules for interconnection" which was mentioned three times. Furthermore, especially the actors in the energy market demand better subsidy schemes, the removal of import restrictions of equipment and more flexibility in the tariff regime. In the following paragraphs, these recommendations will be described in detail.

Loosen import restrictions of equipment intends to reduce the procurement costs, as EnDev's and EDM's representatives explain. Political decision makers should consider, though, that the reduction of import restrictions can interfere with Mozambique's industrialization goals. It can be argued that Mozambique's industries need some protection to develop their competitiveness before their markets are opened for global trade. If immature industries are not protected from more competitive commodities from the global market, their surplus and market share is expected to decrease (Kempa 2012, 160). Eventually this process can lead to the deindustrialization of an economy. Furthermore, it should be considered that already today, Mozambique's import restrictions are not extraordinarily high for typical equipment for rural electrification. For example, current import duties for solar electrification equipment start at 0% for imports from other SADC countries and go up to a maximum of 20% for imports from the rest of the world (DFID 2016, 5).

Instead of lowering import duties, equipment costs can be reduced if taxes are lowered. Correspondingly, Great Britain´s Department for International Development (DFID) urges for a tax holiday for off-grid components. Specifically, DFID demands a 0%-rate for the value added tax on off-grid equipment to boost rural electrification (DFID 2016, 5). Considering that EDM already today benefits from tax and import duty exemptions (ibid.), it could be reasonable to extend these advantages also to the off-grid market. However, political decision makers should take into consideration the tax revenue losses for the public sector which result from such measures.

Apart from reducing taxes or import duties, making regulation more reliable and understandable can be another effective measure to enhance the attractiveness of Mozambique´s energy market. According to FUNAE´s head of department, it should be one of the prior tasks of the government to end the current situation that the regulative and legal system is so unclear that it keeps away new actors. Making regulation easier includes to set clear and transparent rules, how companies can receive permits and concessions. Setting clear and traceable rules is a first step to improve legal security since unclear and unreliable processes are vulnerable to corruption and rent-seeking. Correspondingly, permits and concessions have to be reliable which requires legal security to be improved.

Regarding tariff regulation, experts urge for more flexibility. According to EnDev´s solar energy consultant, differentiated tariffs can be a strong economic incentive for small scale businesses or community projects to pursue electrification. Only if prices are cost-covering, energy projects can be sustainable. Also, differentiated tariffs allow for cross-subsidizing within a company such that revenues from regions with a higher ability to pay can cover the revenue gaps appearing in poorer regions. Furthermore, flexible tariffs in combination with a smart load management reacting to changes in power scarcity, improve efficiency and stabilize the grid system.

Regulation also includes the issue of subsidies. To support rural electrification, donors and government should increase available subsidies for energy projects, FUNAE´s head of department says. For connected mini grids or independent generation capacities, feed-in-tariffs can be an important and effective subsidy scheme. Therefore, representatives from

EnDev and Eduardo Mondlane University urge for a quick introduction of the planned feed-in-tariff regime.

The respondents also recommend improving the non-financial conditions for the connection of mini grids. First and foremost, the respondents demand transparent planning and legal security regarding grid extension and interconnection. For this end, clear rules for interconnection and a grid code that specifies technical requirements are necessary, says EDM´s staff. According to FUNAE´s head of department, the government should allow three possibilities for interconnection:

- Independent operation of the connected mini grid with financial compensations (feed-in-tariffs) for energy fed into the grid,
- Operation of the mini grid on behalf of EDM with a fixed financial compensation,
- Selling the mini grid and handing over operation to EDM.

It is worth to mention that FUNAE´s head of department does not recommend any models for interconnection in which the former mini grid operator and EDM operate the mini grid together. The joint operation of a mini grid can lead to coordination problems and additional transaction costs, such that it can be expected as inefficient and ineffective if one partner engages in distribution and the other one in power generation within the same grid network.

Independently of the specific model, each interconnection requires clear rules and reliable legal processes. Especially for the first option (independent operation with compensation for power fed into the grid), a power purchase agreement is essential. To facilitate such agreements, as mentioned before, five experts explicitly demand a standardization of power purchase agreements. The mini grid researcher at Eduardo Mondlane University additionally recommends specific standardized power purchase agreements for renewables. Looking at renewables, agreements inter alia should address the challenge of fluctuations and special support schemes in order to build an environmentally friendly energy supply. For example, the agreement can prescribe a sufficient level of load management to moderate fluctuations. If the desired level of power stability is not achieved, the agreement can include sanctions. Specific power purchase agreements for renewables can incentivize the implementation of additional and more sustainable generation capacity to tackle energy poverty in Mozambique. Consequently, the standardization of power purchase agreements is a prominent goal for the ministry´s staff, too. As already mentioned, the two respondents from the Ministry of Mineral Resources and Energy say that their authority is already participating in the working group to develop specific standardized power purchase agreements for Mozambique (see chapter 8.4).

Several of the recommendations concern access to information. Respondents request that access to reliable data for the Mozambican energy sector improves. For this end, some of them recommend supporting regional viability studies for different solutions and to acquire specific external consulting to accelerate electrification in Mozambique.

The representatives of Eduardo Mondlane University state that profound and up-to-date studies about the potential of renewable energies in different areas of the country are missing. Together with FUNAE´s head of department, the respondents think that it should be the government´s responsibility to close this knowledge gap by supporting relevant studies. FUNAE´s head of department adds that not only the potentials of renewables should be analyzed but also the viability of different distribution infrastructures (e.g. mini grids). Eduardo Mondlane University´s professor of renewable energy confirms that *"we need specific studies for different regions about which energy solution is viable."*

An important prerequisite for profound feasibility studies and market assessments is the availability of reliable and up-to-date statistical data. Consequently, respondents demand to channel more public resources into collecting and processing energy sector data for Mozambique. For example, better statistics for energy access are requested to determine which regions should be prioritized for electrification. According to FUNAE´s head of department, procuring better data for energy access means that the government should conduct own studies instead of only relying on EDM´s data. The respondent illustrates that in some regions – especially in regions close to South Africa – many households have access to modern energy because they buy solar home systems without participating in a governmental or donor-driven distribution program while these types of energy access and their regional concentration are not counted in the official statistics. Tenenbaum et al. (2014, 324) specify what high-quality electricity sector data should cover. According to the authors, governments – if necessary with the support of donors – should publish data about the dissemination of the electricity market, market size, market segments, customer classes, willingness to pay, regional electrification rates and sources of electricity.

Access to information also includes access to knowledge about the technical aspects of different options for electrification, their proper application, their advantages and disadvantages. Such information can be acquired in vocational training or apprenticeships, taught at vocational schools and universities or it can be acquired from external consulting and training if it is not available inside the country. It can be expected that long term programs for vocational training will be required to generate a general basic level of the necessary technical skills and knowledge in Mozambique. A structural improvement of vocational education will most probably be required to close the current bottleneck of knowledge and skills. Improving access to how-to knowledge and principles knowledge requires furthermore access to such educational programs without prohibitive social or financial barriers. Accordingly, EDM's manager of projects and financing clearly demands to improve basic education which according to the respondent is the foundation for a profound technical education, later in a person's life. For vocational training and universities, EDM's representatives demand implementing more specific energy knowledge into the curricula. According to them, universities and vocational schools in Mozambique offer programs which are very theoretical but include too little practical experience and specific skills for a technical or administrative job in the electricity sector.

In addition to formal education, development cooperation, external consulting and exchange of experience can facilitate the transfer of relevant knowledge about electrification to the Mozambican population, the representatives of the Ministry of Mineral Resources and Energy say. The respondents agree that education and access to knowledge can furthermore increase innovativeness and independent development. Keeping the population informed and involving local communities into electrification projects not only enhances knowledge and supports the development of skills but typically also keeps acceptance high for the corresponding projects (Prindle, Koszalka 2012, 367).

EDM's manager of projects and financing emphasizes that the government's focus should rest on the education of the Mozambican population instead of acquiring external knowledge. According to the respondent, long-term development requires improving the state of education and vocational training inside the country. The respondent states that knowledge of consultants or donor organization can only have a short-term impact since it is exclusively available for a short term and is sometimes insensitive for the local specificities. *"Always, there are consultants from outside the country"*, the respondent says. *"Why not rely more on local expertise and support Mozambican universities?"*.

To support digitalization in the Mozambican electricity sector, the respondents recommend smart improvements of the central grid and more efforts to close remaining gaps in the mobile phone network and mobile internet. Mobile communication is a necessary condition for the viability of off-grid systems. It enables the application of technologies like mobile payment and remote monitoring which are necessary for pay-as-you-go systems. Without such technologies, administration, maintenance and payment collection become prohibitively costly. Therefore, the government – which is responsible for awarding concessions to mobile communications companies – should push the closure of remaining bottlenecks, says the scientific advisor as one of the representatives of the Ministry of Mineral Resources and Energy. Furthermore, public institutions should use digitalization for themselves to make processes easier for potential investors, as the renewables scientist states. For instance, administrative processes to get licenses, permits or general information should be made available to lower the transaction costs connected to entering the Mozambican electricity market.

The respondents´ recommendations offer approaches to improve the framework conditions of smart electrification in Mozambique. To some extent, barriers which are not directly addressed by these recommendations can also be tackled by the proposed measures. For example, more legal security and better regulatory incentives can attract additional investors. Thus, the problem of capital shortness – one of the most important bottlenecks, according to this study – can be relieved. More organizational efficiency, a better coordination and more planning transparency can lead to a faster and more effective implementation of electricity projects. A better digital infrastructure can facilitate the implementation of smart technologies which have the potential to substantially improve Mozambique´s electricity supply, on the grid but especially for off-grid solutions.

10. Conclusion and outlook

In many different ways, this study shed some light on its two essential questions: is there a smart energy supply in Mozambique´s energy future and if so, what does it look like? It is worth to emphasize that the analysis showed clearly that concepts for electrification and smart energy supply from industrialized Western countries cannot simply be transferred to and implemented in a country like Mozambique which is characterized by different and very specific economic, cultural and political framework conditions. Regarding smart electrification, the discussion in industrialized countries is typically focused on grid-based infrastructure. In most cases, industrial countries are supplied by a nationwide comprehensive central grid network. Under such circumstances, the question of smart energy supply addresses upgrading the central grid infrastructure. In Mozambique, however, analyzing potentials for a smart electricity supply requires a view beyond the concept of a conventional central grid with large centralized generation capacities. Already today, many regions in Mozambique are supplied by decentral solutions like mini grids or off-grid systems which in many cases already involve smart technologies. These decentral solutions come with large potentials to expand and to be technologically further refined in the future. That is, smart power supply in Mozambique does also but not mainly concern a smart optimization of the main grid. Instead, such smart solutions which increase the benefits and the convenience of off-grid solutions and mini grids come to the fore.

Broadly summarized, available options for a smart energy supply in Mozambique range from a central smart grid, over isolated and connected mini grids to off-grid systems. It can be clearly stated that generally smart grid solutions and decentral smart energy installations are feasible in Mozambique. In some aspects of digitalization in the energy sector, Mozambique is even among the pioneers. For example, companies in Mozambique offer innovative services for digital payment and distant billing of energy use. With this technology, decentral energy use from all over the country can be billed centrally at one single point of remote control. This exemplary development illustrates how digitalization can support a simultaneous process of decentralization of energy supply and the centralization of control. More specifically speaking, generation and distribution of energy can be decentralized and diversified while administrative processes, such as load management, billing and failure control can be bundled. Thus, efficiency can be increased.

It should be remembered that the different options´ feasibilities vary substantially depending on local conditions. Important determinants which affect central grid infrastructure, mini grids and off-grid systems very differently, are the density of population at a certain site, proximity of central grid infrastructure, local ability and willingness to pay along with the availability of external funds. Each option is viable for different framework conditions. Since Mozambique is quite a diverse country regarding these determinants and framework conditions, it can be inferred that a simultaneous implementation of the available options with a regional differentiation most effectively exploits the specific strengths of each option and makes the most productive use of the scarce available resources. Consequently, it can be expected that Mozambique will most likely experience a parallel implementation of the different options. It shall be stressed that findings also indicate an increasing focus on decentral options such as mini grids and off-grid solutions within the overall expansion of power supply in Mozambique.

What precisely are the specific advantages and disadvantages of the different available options for a smart electrification in Mozambique? While the preceding chapters contained an extensive analysis of this question, this conclusion will only highlight the most important results.

According to the analysis conducted in this study, central grid infrastructure is the best solution for agglomerations of people with a relatively high willingness to pay – criteria typically met in cities and sub-urban areas. On the other hand, extending the central grid to relatively poor areas with a widespread population – conditions given in many parts of Mozambique – is not recommendable. Mozambique is not a pioneer when it comes to the application of smart technologies in the central grid. Despite some first steps to apply digital innovations in order to improve load management, the high costs of upgrading, administrative reluctance and a lack of know-how impede the installation of advanced smart components – especially when it comes to demand response measures. So, to some extent, the central grid can certainly contribute to a better and smarter energy supply but its feasibility is confined to sites with good framework conditions and depends crucially on the perception of smart technologies by essential stakeholders.

Mini grids turn out to be the most feasible option for rather sparsely populated areas without access to the main grid. Nevertheless, a basic level of consumer bundling per unit of grid infrastructure is still necessary for a mini grid to be economically viable. Otherwise the relation between infrastructure costs and paying consumers per unit of infrastructure would be negative, such that the mini grid project would eventually generate losses. Enabling conditions for mini grids are given in a relevant part of the Mozambican territory. Spatial studies and the expert interviews indicate that about one quarter of Mozambique's population would most efficiently be connected to electric energy by this option. For mini grids, especially if fueled by intermittent renewable energies, intelligent load management and monitoring are essential prerequisites for load stability. Consequently, smart energy distribution is likely to continue to diffuse through the Mozambican energy sector.

The interconnection of different sub-systems to a larger grid can be a strong driver for smart energy. At the point of common coupling, measuring the amount of energy exchanged between the sub-systems, balancing loads, failure detection and automated response are important to ensure a stable and smooth grid performance. Connecting several intelligent and autonomous sub-units to each other can furthermore result in a highly diversified, decentrally structured and stable smart grid network. However, there is high evidence that connecting mini grids to the main grid or to each other is much more challenging than installing isolated mini grids in Mozambique. Core prerequisites for a successful interconnection are not or not yet given. For instance, standardized power purchase agreements haven not been fully developed so far, feed-in tariffs still await introduction, a uniform grid code and transparent planning of grid extension do not exist.

Off-grid home systems are identified as the only feasible option for areas characterized by a very low density of population and low income levels. Due to enabling framework conditions, the diffusion of off-grid home systems is expected to continue at a relatively high rate of adoption and even accelerate in the next years. Nevertheless, it should not be neglected that so far, home systems can often not offer the same quality of energy supply as grid connections. Since smart off-grid systems are still quite a new technology, it can be assumed, though, that there are still relevant potentials for improvement and development such that the reliability and supply quality of off-grid electricity could further improve in the coming years.

Generally, technological change and digitalization to improve the energy supply can be expected to be highly welcome to the vast majority of Mozambicans. The categorization of the Mozambican society as a social system with rather materialist value structures indicates that the improvements for physical security, health and comfort coming with a better power supply outweigh possible concerns about data privacy or environmental impacts. A severe scarcity of energy typically leads to a higher appreciation of electrification in comparison to the fulfillment of post material needs such as the protection of personal data.

The relatively small reservations against the use of personal data or the installment of energy infrastructure can be expected to persist in the close future. The presumed rather materialist value structure in Mozambique is presently unlikely to be changed due to the given political and economic framework conditions which are capable to cause insecurity for the generation of Mozambicans who are currently in their formative years. Therefore, the extension of energy infrastructure and the intensive use of information and communication technology in the energy sector can be expected to continuously meet high acceptance levels in the Mozambican society.

Due to the large potentials of smart, sustainable and decentral energy solutions, unelectrified areas in Mozambique can directly benefit from recent technological advancements and renewable energies. That is, besides the possibilities for the successive introduction of smart energy technologies, the Mozambican energy sector can especially benefit from technological leapfrogging. It is worth stressing that this technological leapfrog can also induce a structural leapfrog. While industrial countries are struggling to decentralize and diversify their existing energy sectors, the high feasibility of decentral technologies such as mini grids and home systems can induce a leapfrog, directly towards a highly decentralized and diversified energy supply in developing countries like Mozambique. It should be remembered that the environmental impact of a decentral and diversified energy sector is typically much smaller than of a large-size, fossil-based energy supply. Thus, experience with a decentral and diverse energy sector from Mozambique or similarly structured countries can become valuable information for countries with a mature but conventional, centralized and fossil-based energy supply in the process of transformation.

If decentral solutions continue to achieve the expected high rates of adoption in the next years, additional possibilities for the digitalization of the Mozambican energy sector emerge. Especially decentral solutions make use of recent digital innovations or even depend on smart technologies. Not only smart mini grids can be connected to the main grid or to each other but also smart off-grid solutions can be interconnected to form a small smart mini grid. Consequently, many relatively autonomously working smart sub-systems can form the building blocks of a larger smart energy network which develops from the bottom up, independently from central planning. This bottom-up smart grid approach avoids severe impediments originating from central administration such as corruption, inefficiencies, a lack of competences and a poor coordination among public institutions.

It is worth highlighting that due to the large potentials of smart off-grid systems and decentral mini grids, rural and often poorer areas play a key role in the diffusion of smart energy in Mozambique. It should be kept in mind that there is not a deterministic correlation between poverty and low innovativeness (Kersting 1996). In contrast, the case of decentral smart energy in Mozambique and other African countries shows how rural and poor areas can be pioneers of the adoption of new technologies. With this strong component or rural innovation, smart energy is an example of how rural and urban innovation can mutually reinforce each other. While the application of decentral energy solutions is mainly attractive for rural, sparsely populated areas, the development of smart energy innovations relies on highly innovative, technology-affine, highly qualified and creative social environments which are typically found in urban surroundings (Kersting 2017). Both – development and application – are necessary procedures within each innovation process, such that in the case of decentral smart energy, urban innovation (development and initial point of innovation diffusion) and rural innovation (implementation and re-invention) supplement each other.

Smart energy innovations can induce substantial economic and social changes in the adopting environments such that they can initialize further innovation processes and eventually lead to more development and a higher quality of living. Based on the results of this study, it can be confirmed that smart energy has the potential to significantly improve access to electricity and promote the environmental friendliness of energy supply in Mozambique and similarly structured countries in Africa – an assumption already proposed by Welsch et al. in 2013 as cited in the introduction of this study. It is recommendable to consider smart and decentral energy innovations as central pillars in every electrification strategy for Mozambique and similarly structured countries in order to achieve the United Nations´ energy-related Sustainable Development Goals, namely the access to affordable modern energy, a reliable and sustainable energy supply and climate change mitigation (UN 2015).

Nevertheless, Mozambique still has a long way to go to achieve a thoroughly digitalized energy sector. Due to the several barriers to smart electrification, it remains questionable if smart energy innovations can unfold their entire potential or if at some point the adverse framework conditions will inhibit their further adoption. As summarized in the preliminary conclusions in chapter 6, several challenges limit the optimistic prospects for smart energy in Mozambique. Political reforms can contribute significantly to address these challenges and to shape conditions such that the rate of adoption of smart energy technologies further increases. Quick and effective reforms are regarded as necessary, especially in the following areas: performance of public institutions, regulation and legal security, access to information, political support for digitalization, quality and maintenance, education and qualification. As pressing needs, the interviewed experts and stakeholders name the standardization of power purchase agreements as well as more political and financial support for digital, renewable and decentral energy supply, including the introduction of effective feed-in tariff schemes and flexible tariffs. Furthermore, effective measures against corruption, a better legal framework, more research, more information transparency and an extended coverage with communication technologies are requested. Promoting digital solutions cannot only be a goal but also a measure to achieve political reform. For example, digital payments and remote billing can reduce incentives for corruption, information technologies can facilitate research and enhance transparency. Regarding the general political situation in Mozambique, it is seen as pivotal to improve the integrity of public institutions and to promote a peaceful settlement of political conflicts. A stable, reliable and peaceful environment can substantially accelerate smart electrification in Mozambique.

Further research should develop specific concepts and strategies to put the policy recommendations into practice. Additionally, scientists can support smart electrification by further improving the different energy options and by identifying specific sites for each of the available options, based on an analysis of the local conditions. Thus, more specialized regional feasibility studies can supplement this study, which took a broad and general look at smart electrification in Mozambique. For all kinds of research on this topic, it is seen as essential to rely on experience and know-how from Mozambican stakeholders. A limited scientific perspective from the cultural background of a Western industrialized country can easily fail to understand or misinterpret the local cultural, political and economic conditions in Mozambique. The methods applied in this study aim to give an example for the inclusion of local knowledge and experience.

Which value does a study about transformation processes in Mozambique´s energy sector have for sustainable development in general? It is important to stress that the Sustainable Development Goals must not be regarded as isolated from each other. Many goals interact with and depend on each other. In this very sense, progress in electrification is a necessary condition to achieve many of the other SDGs, such as education-, health- or economy-related goals. Smart solutions and innovative decentral, renewables-based energy supply can contribute substantially to a cleaner, reliable and affordable energy supply. Therefore, it is regarded as appropriate for the United Nations to rank the access to affordable modern energy and a reliable and sustainable energy supply (UN 2015) among the top-priorities for international development. The success of the programs for a sustainable international development will consequently also depend on how effectively the relevant actors make use of the potentials of a smarter energy supply.

References

Abu-Sharkh, S. et al. (2006): Can microgrids make a major contribution to UK energy supply? Renewable and Sustainable Energy Review 10, 78-127.

African Union, AU (2012): List of Countries which have signed, ratified/acceded to the Constitutive Act of the African Union. AU, Addis Ababa.

Ahlborg, H., Hammar, L. (2014): Drivers and barriers to rural electrification in Tanzania and Mozambique – Grid-extension, off-grid and renewable energy technologies. Renewable Energy 61, 117-124.

Almond G. A., Verba, S. (1963): The Civic Culture: Political Attitudes and Democracy in Five Nations. Princeton University Press, Princeton.

Ansprenger, F. (1997): Politische Geschichte Afrikas im 20. Jahrhundert. Beck, München.

Arndt, J. (1967): Role of product-related conversations in the diffusion of a new product. Journal of Marketing Research 4, 291-295.

Associação Lusófona de Energias Renováveis, ALER (2017): Energias Renováveis em Moçambique. Relatório Nacional do Ponto de Situação. ALER, Lisbon.

Associação Lusófona de Energias Renováveis, ALER (2018): Recent Developments in Mozambique's Energy Sector. http://www.aler-renovaveis.org/en/communication/news/recent-developments-in-mozambiques-energy-sector/ [05/20/2019].

Associação Moçambicana de Energias Renováveis, AMER (2019): Sobre nós. http://amer.org.mz/wp_new/ [05/10/2019].

Bass, F. M. (1969): A new product growth model for consumer durables. Management Science 15, 215-227.

Baumol, W. (1977): On the Proper Cost Test for Natural Monopoly in a Multiproduct Industry. The American Economic Review 67, 809-822.

Baumol, W. et al. (1988): Contestable Markets and the Theory of Industry Structure. Harcourt Brace Jovanovich, San Diego.

Behnisch, T. (2010): Mosambikanische Revolte. Hintergründe der Unruhen des 1. und 2. September 2010. Mosambik Rundbrief 81, 7-8.

Benett, C. J. (1992): Regulating Privacy. Data Protection and Public Policy in Europe and the United States. Cornell University Press, New York.

Berg-Schlosser, D., Kersting, N. (1996): Warum weltweite Demokratisierung? Zur Leistungsbilanz demokratischer und autoritärer Regime. In: Hanisch, R.: Demokratieexport in die Länder des Südens? Deutsches Überseeinstitut, Hamburg.

Berry, J. M. (1999): New Liberalism: The Rising Power of Citizen Groups. The Brookings Institution, Washington, DC.

Bewley, T. F. (1995): A depressed labor market as explained by participants. American Economic Review 85, 250–254.

Bewley, T. F. (1999): Why Wages Don't Fall during a Recession. Harvard University Press, Cambridge, MA.

Bewley, T. F. (2002): Interviews as a valid empirical tool in economics. Journal of Socio-Economics 31, 343-353.

Bird, K. et al. (2010): Conflict, education and the intergenerational transmission of poverty in Northern Uganda. Journal of International Development 22, 1183–1196.

Blanchard, O., Illing, G. (2009): Makroökonomie. Pearson, München.

Blyden, B. K., Lee, W.-J. (2006): Modified Microgrid Concept for Rural Electrification in Africa. Power Engineering Society General Meeting, Montreal.

Böhringer, C., Löschel, A. (2005): Climate Policy Beyond Kyoto: Quo Vadis? A Computable General Equilibrium Analysis Based on Expert Judgements. Kyklos 58, 467-493.

Breidert, C. (2006): Estimation of Willingness-to-Pay. Theory, Measurement, Application. Deutscher Universitäts-Verlag, Wiesbaden.

Bugaje I. M. (2006): Renewable energy for sustainable development in Africa: a review. Renewable and Sustainable Energy Reviews 10, 603-612.

Cabrita, João M. (2000): Mozambique. The Tortuous Road to Democracy. Palgrave, New York.

Caldeira, A. (2017). Electricidade volta a ficar mais cara, pelo terceiro ano consecutivo pagam mais os consumidores domésticos em Moçambique. Jornal A Verdade, August 15, 2017. http://www.verdade.co.mz/tema-de-fundo/35-themadefundo/ 63110-electricidade-volta-a-ficar-mais-cara-pelo-terceiro-ano-consecutivo-pagam-mais-os-consumidores-domesticos-em-mocambique- [08/17/2015].

Caldeira, A. (2018 a): "Estamos numa tarifa média de 8 cêntimos por qilowatt/hora, para um custo de 10 cêntimos" EDM sobre sua insustentabilidade. Jornal A Verdade, May 09, 2018. http://www.verdade.co.mz/tema-de-fundo/35-themadefundo/65704-estamos-numa-tarifa-media-de-8-centimos-por-quilowatthora-para-um-custo-de-10-centimos-edm-sobre-sua-insustentabilidade [14/05/2018].

Caldeira, A. (2018 b): IDE vai crescer em 2019 com início dos investimentos do gás natural em Moçambique. Jornal A Verdade, October 05, 2018. http://www.verdade.co.mz/nacional/67018-ide-vai-crescer-em-2019-com-inicio-dos-investimentos-do-gas-natural-em-mocambique [10/16/2018].

Carus, A. W., Ogilvie, S. (2009): Turning qualitative into quantitative evidence: a well-used method made explicit. Economic History Review 62, 893-925.

Central Intelligence Agency, CIA (2016): The World Factbook. Population. https://www.cia.gov/library/publications/the-world-factbook/fields/2119.html [04/08/2016].

Cescon, G. (2015): Devergy solar pre-paid micro-grids for rural villages in Tanzania. Smart Grids Conference Proceedings 2015, African Utility Week, Cape Town.

Chilton, S. M. and Hutchinson, W. G. (2003): A qualitative examination of how respondents in a contingent valuation study rationalise their WTP responses to an increase in the quantity of the environmental good. Journal of Economic Psychology 24, 65–75.

Chongo Cuamba, B. et al. (2006): A solar resources assessment in Mozambique. Journal of Energy in Southern Africa 17, 76-85.

Cipriano, A. et al. (2015): The Electricity Sector in Mozambique. An Analysis of the Power Crisis and its Impact on the Business Environment. USAID, Maputo.

Clark, J. et al. (2000): I struggled with this money business: respondents' perspectives on contingent valuation. Ecological Economics 33, 45–62.

Coase, R. (1937): The Nature of the Firm. Economica 4, 386-405.

Cobb, C. W., Douglas, P. H. (1928): A Theory of Production. The American Economic Review 18, 139-165.

Collier, P. (2008): Die unterste Milliarde. Warum die ärmsten Länder scheitern und was man dagegen tun kann. C.H. Beck, München.

Cournot, A. (1924): Untersuchungen über die mathematischen Grundlagen der Theorie des Reichtums. Fischer, Jena.

Couture, T., Gagnon, Y. (2010): An analysis of feed-in tariff remuneration models: Implications for renewable energy investment. Energy Policy 38, 955-965.

Crispim, J. et al. (2014): Smart Grids in the EU with smart regulation: Experiences for the UK, Italy and Portugal. Utilities Policy 31, 85-93.

Dada, J. O. (2014): Towards understanding the benefits and challenges of Smart/Micro-Grid for electricity supply system in Nigeria. Renewable and Sustainable Energy Reviews 38, 1003-1014.

Dalton, R. (1977): Was There a Revolution? A Note on Generational versus Life Cycle Explanations of Value Differences. Comparative Political Studies 9, 459-473.

Department for International Development, DFID (2016): Energy Africa – Mozambique. Technical Assistance to model and analyse the economic effects of VAT and tariffs on picoPV products, Solar Home Systems and Improved Cookstoves. DFID, London.

Desaigues, B. (2001): Is expressed WTP consistent with welfare economics? A response from 73 cognitive interviews. Swiss Journal of Economics and Statistics 137, 35–47.

Deutsche Gesellschaft für International Zusammenarbeit, GIZ (2018): About GIZ. Identity. https://www.giz.de/en/aboutgiz/identity.html [05/11/2018].

Djankov, S. et al. (2008): The curse of aid. Journal of Economic Growth 13, 169-194.

Dogson, M. (1993): Technological Collaboration in Industry. Routledge, London.

Dorussen, H. et al. (2005): Assessing the Reliability and Validity of Expert Interviews. European Union Politics 6, 315-337.

Douglas, P. H. (1976): The Cobb-Douglas Production Function Once Again: Its History, Its Testing, and Some New Empirical Values. Journal of Political Economy 84, 903-916.

Electric Power Research Institute, EPRI (2011): Estimating the Costs and Benefits of the Smart Grid. Preliminary Estimate of the Investment Requirements and the Resultant Benefits of a Fully Functioning Smart Grid. EPRI, Palo Alto.

Electricidade de Moçambique, EDM (2013): Relatório Annual de Estatística. EDM, Maputo.

Electricidade de Moçambique, EDM (2016): Tarifarios de Energia Eléctrica. http://www.edm.co.mz/index.php?option=com_content&view=article&id=121&Ite mid=83&lang=pt [04/05/2016].

Electricidade de Moçambique, EDM (2017): Tarifarios de Energia Eléctrica. http://www.edm.co.mz/index.php?option=com_content&view=article&id=121&Ite mid=83&lang=pt [10/05/2017].

Electricidade de Moçambique, EDM (2018): Tarifarios de Energia Eléctrica. http://www.edm.co.mz/index.php?option=com_content&view=article&id=121&Ite mid=83&lang=pt [11/22/2018].

Emerson, Stephen A. (2014): The Battle for Mozambique: The Frelimo-Renamo Struggle, 1977-1992. South Publishers, South Africa.

Endres, A. (2013): Umweltökonomie. Kohlhammer, Stuttgart.

Energising Development, EnDev (2019): About Energising Development. htpps://endev.info/content/Main_Page [03/19/2019].

Ernst & Young (2012): Smart Grid: a race worth winning? A report on the economic benefits of a smart grid. SmartGridGB, Ernst & Young, London.

Eucken, W. (1965): Grundsätze der Wirtschaftspolitik. Rowohlt/Mohr, Tübingen.

Eveland, J. D. et al. (1977): The Innovative Process in Public Organizations. University of Michigan, Ann Arbor.

Fabricius, P. (2016): Mozambique: devolution or revolution? ISS Today, Institute for Security Studies. https://www.issafrica.org/iss-today/mozambique-devolution-or-revolution [11/24/2016].

Faust, J., Leiderer, S. (2008): Zur Effektivität und politischen Ökonomie der Entwicklungszusammenarbeit. Politische Vierteljahresschrift 49, 129-152.

Fiol, C. M., Lyles, M. A. (1985): Organisational Learning. Academy of Management Review, 803-813.

Flick, U. (1999): Qualitative Forschung. Theorie, Methoden, Anwendung in Psychologie und Sozialwissenschaften. Rowohlt, Reinbeck.

Folha de Maputo (2016): Mozambique will reduce inflation to 14% in 2017. http://clubofmozambique.com/news/mozambique-will-reduce-inflation-14-percent-2017-central-bank/ [08/18/2017]

Frank, R. E. et al. (1964): The determinants of innovative behavior with respect to a branded, frequently purchased food product. In: Smith, L. G.: Proceedings of the American Marketing Association. American Marketing Association, Chicago.

Fritsch, M. (2011): Marktversagen und Wirtschaftspolitik. Mikroökonomische Grundlagen staatlichen Handelns. Vahlen, München.

Fritsch, M. (2014): Marktversagen und Wirtschaftspolitik. Mikroökonomische Grundlagen staatlichen Handelns. Vahlen, München.

Frontline (2017): The FrontlineSMS Platform. http://www.frontlinesms.com/product/ [08/17/2017].

Fudenberg, D. et al. (1983): Preemption, leapfrogging, and competition in patent races. European Economic Review 22, 3-31.

Fundo de Energia, FUNAE (2014): Atlas das Energias Renaváveis de Moçambique. FUNAE, Maputo.

Giesbert, L., Schindler, K. (2012): Assets, Shocks, and Poverty Traps in Rural Mozambique. World Development 40, 1594-1609.

Gossen, H. H. (1854): Entwicklung der Gesetze des menschlichen Verkehrs, und der daraus fließenden Regeln für menschliches Handeln. Vieweg, Braunschweig.

Government of Mozambique (2007): Decree 48/2007.

Government of Mozambique (2009): Resolução 10/2009 de 4 de Junho: Estratégia do Sector de Energia. Boletim da República. I Série, Número 22, 1-22.

Government of Mozambique (2011): Law 15/2011.

Government of Mozambique (2014): Decree 58/2014.

Government of Mozambique (2015): Proposta do Programa Quinquenal do Governo 2015-2019. Governo da República de Moçambique, Maputo.

Green, L. W. (1986): The Theory of Participation. A Qualitative Analysis of Its Expression in National and International Health Policies. Advances in Health Education and Promotion 1, 211-236.

GreenLight Consult (2016): Field Report: Market Attractiveness Analysis and Demand Assessment for M-Kopa Solar Systems in Mozambique. GreenLight Consult, Maputo.

Gugler, K. et al. (2013): Ownership unbundling and investment in electricity markets – A cross country study. Energy Economics 40, 702-713.

Hamel G., Prahalad, C. K. (1991): Corporate imagination and expeditionary marketing. Harvard Business School Review 69, 81-92.

Hamilton, B. et al. (2012): The Customer Side of the Meter. Sioshansi, F. P.: Smart Grid. Integrating Renewable, Distributed and Efficient Energy. Academic Press, Waltham, 397-418.

Hartwig, K. H. (2004): Europäische Airline Industrie im Trade Off. In: Fritsch, M.: Marktdynamik und Innovation. Gedächtnisschrift an Hans-Jürgen Ewers. Duncker und Humblot, Berlin.

Hassinger, E. (1959): Stages in the Adoption Process. Rural Sociology 24, 52-53.

Hedberg, B. L. (1989): The Age of Unreason. Business Books, London.

Heinze, T. (2001): Qualitative Sozialforschung. Einführung, Methodologie und Forschungspraxis. Oldenbourg, München.

Hölmstrom, B. (1979): Moral Hazard and Observability. The Bell Journal of Economics 10, 74-91.

Human Rights Watch (2019): Mozambique. Events of 2018. https://www.hrw.org/world-report/2019/country-chapters/mozambique [04/27/2019].

Igbaria, M. et al. (1994): The Retrospective Roles of Perceived Usefulness and Perceived Fun in the Acceptance of Microcomputer Technology. Behavior and Information Technology 13, 349-361.

Inglehart, R. F. (1977): The Silent Revolution. Changing Values and Political Styles Among Western Publics. Princeton University Press, Princeton.

Inglehart, R. F. (2008): Changing Values among Western Publics from 1970 to 2006. West European Politics 31, 130-146.

Inglehart, R. F., Flanagan, S. C. (1987): Value Change in Industrial Societies. The American Political Science Review 81, 1289-1319.

Instituto Nacional de Estatística, INE (2014): Estatísticas dos Transportes e Comunicações. INE, Maputo.

Instituto Nacional de Estatística, INE (2016): Contas Nacionais. 2° Trimestre 2016. INE, Maputo.

International Energy Agency, IEA (2012): Energy Technology Perspectives 2012. Pathways to a Clean Energy System. Executive Summary. IEA, Paris.

International Energy Agency, IEA (2017): Energy Access Outlook: World Energy Outlook 2017 Special Report. Electricity Access Database. IEA, Paris.

International Energy Agency, IEA (2018): Methodology. Defining energy access. http://www.iea.org/energyaccess/methodology/ [09/03/2018].

International Monetary Fund, IMF (2016): IMF Staff Concludes Visit to Mozambique. Press release. http://www.imf.org/external/np/sec/pr/2016/pr16304.htm [09/02/2016].

International Monetary Fund, IMF (2018): GDP, current prices. https://www.imf.org/external/datamapper/NGDPD@WEO/OEMDC/ADVEC/WEO WORLD [11/15/2018].

International Telecommunication Union, ITU (2019): Statistics. https://www.itu.int/en/ITU-D/Statistics/Pages/stat/default.aspx [03/26/2019].

Jehle, A. J., Reny P. J. (2011): Advanced Microeconomic Theory. Prentice Hall, Essex.

Jones, S. (2010): The economic contribution of tourism in Mozambique: Insights from a Social Accounting Matrix. Development Southern Africa 27, 679-696.shar

Jornal Notícias (2017): Dhlakama anuncia trégua por tempo indeterminado. http://www.jornalnoticias.co.mz/index.php/politica/67199-dhlakama-anuncia-treguas-indeterminadas.html [18/05/2017].

Kaiser, F. J., Kaminski, H. (2012): Methodik des Ökonomieunterrichts. Klinkhardt, Bad Heilbrunn.

Karekezi, S. (2002): Renewables in Africa – meeting the energy needs of the poor. Energy Policy 30, 1059-1069.

Kelly et al. (2000): Bridging the Gap Between the Science and Service of HIV Prevention: Transferring Effective Research-Based HIV Prevention Interventions to

Community AIDS Service Providers. American Journal of Public Health 90, 1082-1088.

Kempa, B. (2012): Internationale Ökonomie. Kohlhammer, Stuttgart.

Kendall, M. G., Babington Smith B. (1939): The Problem of m Rankings. Annals of Mathematical Statistics 10, 275-287.

Kendall, M.G. (1948). Rank correlation methods. Griffin, London

Kersting, N. (1994): Demokratie und Armut in Zimbabwe. Politische Partizipation und urbaner Lebensstil. LIT Verlag, Münster.

Kersting, N. (1996): Urbane Armut. Überlebensstrategien in der „Dritten Welt". Verlag für Entwicklungspolitik, Saarbrücken.

Kersting, N. (2013): Online participation: From `invited´ to `invented´ spaces. International Journal of Electronic Governance 6, 270-280.

Kersting, N. (2017): Urbane Innovation – Ursachen, Strategien und Qualitätskriterien. In: Kersting, N.: Urbane Innovation. Springer VS, Wiesbaden.

Kersting, N. et al. (2009): Local Governance Reform in Global Perspective. VS Verlag, Wiesbaden.

Kersting, N., Matsiko, A. (2018): The State of E-local Participation in Kampala Capital City Authority in Uganda: A Reality or Deception? In: Edelmann, N. et al.: Electronic Participation. ePart 2018. Lecture Notes in Computer Science 11021, 76-88.

Kersting, N., Sperberg, J. (2003): Political Participation. In: Berg-Schlosser, D., Kersting, N.: Poverty and Democracy. Zed Books, New York, 153-180.

Kersting, N., Zhu Y. (2018): Crowd Sourced Monitoring in Smart Cities in the United Kingdom. In: Alexandrov, D. A.: Digital Transformation and Global Society. Third International Conference, DTGS 2018, Revised Selected Papers, Part I. Springer, Cham, 255-266.

Kevenhörster (2009): Entwicklungspolitik. VS-Verlag, Wiesbaden.

Kevenhörster, P. (2014): Entwicklungshilfe auf dem Prüfstand. Entwicklungspolitische Bilanzen führender Geberstaaten. Waxmann, Münster.

King, C. W. (1963): Fashion adoption: A rebuttal to the trickle down theory. In: Greyser, S. A.: Proceedings of the American Marketing Association. American Marketing Association, Chicago.

Klemp, L., Poeschke, R. (2005): Good Governance gegen Armut und Staatsversagen. Aus Politik und Zeitgeschichte 28-29, 16-25.

Knutson, J. N. (1972): The Human Basis of the Polity: A Psychological Study of Political Men. Aldine-Atherton, Chicago.

Kromrey, H. (2002): Empirische Sozialforschung – Modelle und Methoden der standardisierten Datenerhebung und Datenauswertung. Leske und Budrich, Opladen.

Kromrey, H. (2009): Empirische Sozialforschung. Modelle und Methoden der standardisierten Datenerhebung und Datenauswertung. Lucius & Lucius, Stuttgart.

Lamnek, S. (2010): Qualitative Sozialforschung. Beltz Psychologie Verlags Union, Weinheim.

Larsen, J. K., Agrawal-Rogers, R. (1977): Re-Invention of Innovation: A Study of Community Health Centers. American Institute for Research in the Behavioral Sciences, Palo Alto.

Lerner, J., Tirole, J. (2000): Some simple economics of open source. Journal of Industrial Economics 50, 197–234.

Lewis, O. (1966): The Culture of Poverty. Scientific American 215, 19-25.

Machena, Y., Maposa, S. (2013): Zambezi Basin Dam Boom Threatens Delta. World Rivers Review 28, 6.

Mahajan V. et al. (1991) New Product Diffusion Models in Marketing: A Review and Directions for Research. In: Nakićenović N., Grübler A.: Diffusion of Technologies and Social Behavior. Springer, Berlin.

Mandlate, F. (2018): Morreu Afonso Dhlakama. O País, May 3, 2018, http://opais.sapo.mz/morreu-afonso-dhlakama [05/11/2018].

Mankiw, G., Taylor, M. (2008): Grundzüge der Volkswirtschaftslehre. Schäffer-Poeschel, Stuttgart.

March J. G. (1981): Footnotes to Organizational Change. Administrative Science Quarterly 26, 563-577.

Mayer, H. O. (2008): Interview und schriftliche Befragung. Oldenbourg, München.

Meredith, M. (2011): The State of Africa. A History of the Continent Since Independence. Simon and Schuster, London.

Meuser, M., Nagel, U. (1991): Experteninterviews – vielfach erprobt, wenig bedacht. Ein Beitrag zur qualitativen Methodendiskussion. In: Garz, D., Kraimer, K.: Qualitativ-empirische Sozialforschung, Springer, Wiesbaden, 441-468.

Meyer, U. et al. (2007): Grundzüge der mikroökonomischen Theorie. Springer, Berlin.

Ministério de Economia e Finanças da República de Moçambique (2016): Probreza e Bem-Estar em Moçambique: Quarta Avaliação nacional. Direcção de Estudos Económicos e Financeiros, Maputo.

Ministério de Planificação e Desinvolvimento, MPD (2010): Probreza e Bem-Estar em Moçambique: Terceira Avaliação Nacional. MPD, Maputo.

Murphy, J. T. (2001): Making the Energy transition in rural East Africa: Is leapfrogging an alternative? Technological Forecasting and Social Change 68, 173-193.

Myers (2013): Qualitative Research in Business and Management. Sage, London.

Myers, M. D., Newman, M. (2007): The qualitative interview in IS research: examining the craft. Information and Organization 17, 2-26.

O´Keefe, T. (2002): Organisational learning: a new perspective. Journal of European Industrial Training 26, 130-141.

Olson, M. (1982): The Rise and Decline of Nations: Economic Growth, Stagflation, and Social Rigidities. New Haven, Yale University Press.

Organization for Economic Co-Operation and Development (2018): Development Aid at a Glance. Statistics by Region. 2. Africa, 2018 Edition. OECD, Paris.

Organization for Economic Co-operation and Development, OECD (2013) Die OECD in Zahlen und Fakten 2013. Wirtschaft, Umwelt, Gesellschaft. OECD, Paris.

Organization for Economic Co-operation and Development, OECD (2016 a): Development Aid at a Glance. Statistics by Region. 2. Africa. 2016 Edition. http://www.oecd.org/dac/stats/documentupload/2%20Africa%20-%20Development %20Aid%20at%20a%20Glance%202016.pdf [04/08/2016].

Organization for Economic Co-operation and Development, OECD (2016 b): Aid at a Glance: Mozambique. https://public.tableau.com/views/OECDDACAidataglanceby recipi-
ent_new/Recipients?:embed=y&:display_count=yes&:showTabs=y&:toolbar=no? &:showVizHome=no [04/08/2016].

Organization of African Unity, OAU (1963): OAU Charter. OAU, Addis Ababa.

Painuly, J. P. (2001): Barriers to renewable energy penetration; a framework for analysis. Renewable Energy 24, 73-89.

Platt, G. et al. (2012): What Role for Microgrids? In: Sioshansi, F. P.: Smart Grid. Integrating Renewable, Distributed and Efficient Energy. Academic Press, Waltham, 185-209.

Popper, K. (1989): Logik der Forschung. Mohr-Siebeck, Tübingen.

Prindle, W., Koszalka, M. (2012): Succeeding in the Smart Grid Space by Listening to Consumers and Stakeholders. In: Sioshansi, F. P.: Smart Grid. Integrating Renewable, Distributed and Efficient Energy. Academic Press, Waltham, 343-371.

Rao, V. R. (2009): Handbook of Pricing Research in Marketing. Edward Elgar, Cheltenham.

Rathnayaka, A.J. et al. (2011): Identifying prosumer's energy sharing behaviours for forming optimal prosumer-communities. IEEE International Conference on Cloud and Service Computing (CSC), Dec 12-14 2011. IEEE, Hong Kong.

Republic of South Africa, RSA (2008): The Energy Act, No. 34, 2008. Pretoria, South Africa.

Ricardo, D. (1817): On the principles of political economy and taxation. Murray, London.

Robertson, T. S. (1967): Determination of innovative behavior. In: Moyer, R.: Proceedings of the American Marketing Association. American Marketing Association, Chicago.

Robinson, J. (1932): Imperfect Competition and Falling Supply Price. The Economic Journal 168, 544-554.

Rogers, E. M. (2003): Diffusion of Innovations. Free Press, New York.

Rokeach, M. (1969): Beliefs, Attitudes and Values. Jossey-Bass, San Francisco.

Rokeach, M. (1973): The Nature of Human Values. Free Press, New York.

Rüttgers, C. (2009): Wettbewerb in der deutschen Trinkwasserwirtschaft? Ein disaggregierter Regulierungsansatz und seine wettbewerblichen Implikationen. Duncker und Humblot, Berlin.

Ryan, B., Gross, N. C. (1943): The Diffusion of Hybrid Seed Corn in Two Iowa Communities. Rural Sociology 8, 15-24.

Sachs, J. D. (2004): Ending Africa´s Poverty Trap. Brookings Papers on Economic Activity 1, 177-240.

Sachs, J. D., Warner, A. M. (1999): The big push, natural resource booms and growth. Journal of development economics 59, 43-76.

Sampablo, M. et al. (2017): Mini Grid Market Opportunity Assessment: Mozambique. African Development Bank, Abidjan.

Samuelson, P. A. (1937): A Note on Measurement of Utility. Review of Economic Studies 4: 155-161.

Schein, E. H. (1990): Organizational Culture. American Psychologist 45, 109-119.

Schreiber et al. (2015): Flexible electricity tariffs: Power and energy price signals designed for a smarter grid. Energy 93, 2568-2581.

Schumpeter, J. A. (1912): Theorie der wirtschaftlichen Entwicklung. Berlin.

Schumpeter, J. A. (2005): Kapitalismus, Sozialismus und Demokratie. UTB, Stuttgart.

Schweitzer Engineering Laboratories, SEL (2017): Programmable Automation Controller. https://selinc.com/products/2411/ [11/08/2017].

Senge, P. M. (2006): The Fifth Discipline. The Art and Practice of the Learning Organization. Curreny Doubleday, New York.

Sharkey, W. (1982): The theory of natural monopoly. Cambridge University Press, Cambridge.

Sharkey, W. (1982): The theory of natural monopoly. Cambridge University Press, Cambridge.

Shayo, D. P., Kersting N. (2016): An Examination of Online Electoral Campaigning in Tanzania. In: Edelmann Noella, Parycek Peter: Proceedings of the International Conference for E-democracy and Open Government. Los Alamitos, California: IEEE Computer Society, S. 69-76.

Shayo, D. P., Kersting, N. (2017): Crowd monitoring in elections through ICT. Krems_Cedem 17. CPS IEEE Computer society, 36-45.

Silk, A. J. (1966): Overlap among self-designated opinion leaders: A study of selected dental products and services. Journal of Marketing Research 3, 255-259.

Sioshansi, F. P. (2012): Smart Grid. Integrating Renewable, Distributed, and Efficient Energy. Academic Press, Waltham.

Smertnik, H. (2015): Mobile for smart solutions: How mobile can help improve energy access in Africa. Smart Grids Conference Proceedings 2015, African Utility Week, Cape Town.

SolarWorks (2018 a): Solar Home System. http://solar-works.co.za/products/solar-home-system#features [04/01/2018].

SolarWorks (2018 b): Integrated off-grid power solutions. http://solar-works. co.za/about-solarworks [04/01/2018].

South African Smart Grid Initiative, SASGI (2016): The South African Smart Grid Initiative. http://www.sasgi.org.za/about-sasgi/ [04/05/2016].

Southern African Development Community, SADC (2015): Consolidated Text of the Treaty of the Southern African Development Community. SADC, Garborone.

Srinivasan, S. (2014): FrontlineSMS, Mobile-for-Development, and the "Long Tail" of Governance. In: Livingston, S., Walter-Drop, G.: Bits and Atoms. Information and Communication Technology in Areas of Limited Statehood. Oxford University Press, New York, 79-97.

Starr, M. A. (2014): Qualitative and Mixed Methods Research in Economics: Surprising Growth, Promising Future. Journal of Economic Surveys 28, 238–264

Steiner, P. O. (1957): Peak loads and efficient pricing. Quarterly Journal of Economics 71, 585-610.

Szabó et al. (2011): Energy solutions in rural Africa: mapping electrification costs of distributed solar and diesel generation versus grid extension. Environmental Research Letters 6, 1-9.

Tenenbaum, B. et al. (2014): From the bottom up. How small Power Producers and Mini-Grids Can Deliver Electrification and Renewable Energy in Africa. The World Bank, Washington D.C..

The Economist (1977): The Dutch Disease. The Economist, November 1977, 82-83.

Tullock, G. (1967): The Welfare Costs of Tariffs, Monopolies, and Theft. Western Economic Journal 5, 224-232.

UNICEF (2018): Mozambique. Statistics. https://www.unicef.org/infobycountry/ mozambique_statistics.html [09/25/2018].

United Nations Conference on Trade and Development, UNCTAD (2016 a): Foreign direct investment: Inward and outward flows and stock, annual, 1980-2014. http://unctadstat.unctad.org/wds/TableViewer/tableView.aspx?ReportId=96740 [04/28/2016].

United Nations Conference on Trade and Development, UNCTAD (2016 b): Personal remittances: receipts and payments, annual, 1980-2014. http://unctadstat. unctad.org/wds/TableViewer/tableView.aspx [04/28/2016].

United Nations Conference on Trade and Development, UNCTAD (2018): Personal remittances: receipts and payments, annual, 1980-2017. http://unctadstat.unctad.org/wds/TableViewer/tableView.aspx [11/15/2019].

United Nations, UN (1992): General Peace Agreement for Mozambique. UN, Rome.

United Nations, UN (1997): Kyoto Protocol to the United Nations Framework Convention on Climate Change. Kyoto, Japan.

United Nations, UN (2015): Transforming our world: the 2030 Agenda for Sustainable Development. United Nations Resolution A/RES/70/1 of 25 September 2015. UN, New York.

United Nations, UN (2016 a): GDP growth. http://data.un.org/Data.aspx?q=GDP+ growth&d=WDI&f=Indicator_Code%3aNY.GDP.MKTP.KD.ZG [04/07/2016].

United Nations, UN (2016 b): Per capita GDP at current prices. http://data.un.org/Data.aspx?q=gdp+per+capita&d=SNAAMA&f=grID%3a101%3 bcurrID%3aUSD%3bpcFlag%3a1 [05/06/2016]

United Nations, UN (2016 c): UN Data. Mozambique. http://data.un.org/ CountryProfile.aspx?crName=Mozambique#Economic [05/24/2016].

United Nations, UN (2016 d): GDP at market prices. Mozambique. http://data.un.org/Data.aspx?q=gdp++&d=WDI&f=Indicator_Code%3aNY.GDP.M KTP.CD [11/10/16].

United Nations, UN (2016): Paris Agreement under the United Nations Framework Convention on Climate Change. Paris, France.

United Nations, UN (2018): UN Data. Mozambique. http://data.un.org/en/iso/mz.html [11/15/2018].

United States Agency for International Development, USAID (2013): Mozambique Climate Vulnerability Profile. USAID, Washington D.C.

Valente, C. (2011): Household returns to land transfers in South Africa: a Q-squared analysis. Journal of Development Studies 47, 354–376.

Venables, A. J. (2016): Using Natural Resources for Development: Why Has It Proven So Difficult? Journal of Economic Perspectives 30, 161-184.

Veremachi et al. (2016): PCM Heat Storage Charged with a Double-Reflector Solar System. Journal of Solar Energy 2016, 1-8.

Victron Energy (2018): Pay-As-You-Go. Solar home systems and mini grids. https://www.victronenergy.de/markets/payg [06/20/2018].

von Lucke, J. (2017): Technische Innovation – Potenziale von Open Government, offenen Daten und intelligenten Städten. In: Kersting, N.: Urbane Innovation. Springer VS, Wiesbaden.

Welsch, M. et al. (2013): Smart and Just Grids for sub-Saharan Africa: Exploring options. Renewable and Sustainable Energy Reviews 20, 336-352.

Wessels, J. (2016): In Mozambique, a return to the horrors of civil war. Dailymail Online. http://www.dailymail.co.uk/wires/afp/article-3619281/In-Mozambique-return-horrors-civil-war.html [09/29/2017].

Williamson, E. (1989): Transaction Cost Economics. In: Schmalensee, R., Willig, R.: Handbook of Industrial Organisation. North-Holland, Amsterdam.

Williamson, O. (1985): The Economic Institutions of Capitalism. Firms, Markets, Relational Contracting. Free Press, New York.

World Bank (2014): Mobile at The Base of the Pyramid: Mozambique. World Bank, Washington D.C.

World Bank (2015): Republic of Mozambique. Mozambique Energy Sector Policy Note. World Bank, Washington.

World Bank (2016 a): World Development Indicators. Inflation, consumer prices, annually 2010-2015. http://databank.worldbank.org/data/reports.aspx?source=2& series=FP.CPI.TOTL.ZG&country=MOZ [11/10/2016].

World Bank (2016 b): Urban population in per cent of total. Mozambique. http://data.worldbank.org/indicator/SP.URB.TOTL.IN.ZS?locations=MZ [11/10/2016].

World Bank (2016 c): Doing Business 2016. Measuring Regulatory Quality and Efficiency. World Bank, Washington.

World Bank (2018 a): Project Information - HMNK Mphanda Nkuwa HPP. Private Participation in Renewable Energy Database. http://ppi-re.worldbank.org/Data/ Project/hmnk-mphanda-nkuwa-hpp-6483 [03/09/2018].

World Bank (2018 b): Doing Business 2018. Reforming to Create Jobs. World Bank, Washington.

World Bank (2018 c): Doing Business. Economy Rankings. http://www.doingbusiness .org/rankings?region=sub-saharan-africa [03/09/2018].

World Bank (2018 d): Population density (people per sq. km of land area). https://data.worldbank.org/indicator/EN.POP.DNST [05/08/2018].

World Bank (2018 e): Key features of a Power and Energy Purchase Agreement (PPA). https://ppp.worldbank.org/public-private-partnership/sector/energy/energy-power-agreements/power-purchase-agreements#key_features [06/07/2018].

World Value Survey (2017): Findings and Insights. http://www.worldvaluessurvey. org/WVSContents.jsp [03/02/2018].

A. Annex

A.1. Questionnaire

Please answer for each variable the following question: In Mozambique, what is the occurrence of...

Variables	Very low						Very high	Don't know
	-3	-2	-1	0	1	2	3	
Available capital for investments in general?	☐	☐	☐	☐	☐	☐	☐	☐
Willingness to spend available capital on smart grid investments?	☐	☐	☐	☐	☐	☐	☐	☐
Competition in the electricity market?	☐	☐	☐	☐	☐	☐	☐	☐
Transaction costs induced by smart grid implementation?[34]	☐	☐	☐	☐	☐	☐	☐	☐
Incentives, induced by power tariffs, to implement smart grid-solutions?[35]	☐	☐	☐	☐	☐	☐	☐	☐
Smart grid-supportive governance?[36]	☐	☐	☐	☐	☐	☐	☐	☐
Smart grid-supportive regulations?	☐	☐	☐	☐	☐	☐	☐	☐
Support of smart grid-implementation by donor involvement?	☐	☐	☐	☐	☐	☐	☐	☐
International coopera-	☐	☐	☐	☐	☐	☐	☐	☐

---

34 Transaction costs are costs for using a market such as information costs, legal consulting, negotiating and enforcing contracts.
35 E.g. incentives for investments in demand-side response, purchasing smart meters inter alia
36 That is, how strongly does the performance of decision-makers, such as government, administration, parliament or political parties facilitate smart grid-implementation in Mozambique?

tion regarding the implementation of new technologies?								
Technical applicability of smart grid infrastructure in Mozambique?	☐	☐	☐	☐	☐	☐	☐	☐
Expected quality of grid management (maintenance and operation) in Mozambique?	☐	☐	☐	☐	☐	☐	☐	☐
Stakeholder[37] acceptance of a smart grid infrastructure in Mozambique?	☐	☐	☐	☐	☐	☐	☐	☐

Notes:

The questionnaire asked the respondents to answer the following question for each potentially relevant variable x on scales from minus three to plus three, including zero: "In Mozambique, what is the occurrence of x?". For each variable, the question was formulated the same way in order to avoid biased answers, arising from different interpretations of varying questions. If a respondent chooses zero on the scale, the interpretation is that the variable x is neither a Mozambique specific driver nor barrier. This answer is of valuable information, reflecting why a scale, containing zero in the middle was applied.

The intervals from zero to minus three and from zero to plus three do not have a middle. Thus, a tendency to the middle in these intervals is avoided. The goal is to get rather clear statements from the experts of whether a variable is a driver or a barrier. A scale with seven increments is sufficient for distinguishing different drivers and barriers and simultaneously brief enough for respondents to ascribe empirical relations to the numerical ones.[38]

In order to avoid uninformed guessing, the answer option "don´t know" was included. If an expert chose "don´t know", he or she excluded himself or herself from the group of experts for the question under consideration.

---

37 Consumers, non-governmental organizations and suppliers inter alia.
38 Basic logical prerequisites for scales, which are met by the proposed one can be found in Kromrey (2002).

Since the interest in this study are the assessments of experts, "don´t know-answers" can be excluded from the sample.

Since closed questions present a finite domain of answer options, there is a risk that relevant drivers or barriers are not included (Kromrey 2002). Therefore, the questionnaire was extended by the following question: *"Are there variables which influence smart grid implementation in Mozambique that are not included in the questionnaire? Please name the neglected variables and specify if they are rather a driver or a barrier to smart grids in Mozambique."* Answers to this question were included in the qualitative evaluation.

Neglected drivers and barriers cannot be added to the set of closed questions later on. Doing so would harm the comparability of results. In order to reduce the risk of neglecting relevant drivers or barriers in the set of closed questions, the detailed theoretical analysis of potential drivers and barriers from the preceding chapters has been completed before phrasing the questions.

The questionnaire can be applied separately from the qualitative interviews. Thus, the number of respondents for the questionnaire can be increased. Experts who answered the questionnaire and participated in an interview were given the questionnaire at first. Although the preceding closed questions in the questionnaire could confine the answers to the open questions in the interviews, it is regarded as reasonable to ask the closed questions before the open ones. Thus, comparability between the answers to the questionnaire by respondents with and without participation in the interviews is ensured. This enhanced comparability – especially important for the output of the questionnaire – is assumed to outweigh a potential influence on the responses to the open questions in the interviews.

A.2. Interview structure

Interview guide for part one of the study (drivers and barriers)

1. Introduction
   - Thank the respondent for his or her participation
   - Ask if respondent accepts to be recorded
   - Ask if the interview can be published with the respondent´s full name or if he or she wishes to remain anonymous
2. General information
   - In what context is your enterprise/institution/organization dealing with smart grids?
   - Can you describe your position in your enterprise/organization/institution?
   - In what context are you personally involved in these projects or programs involving smart grids?
3. Finance
   - How would you describe the availability of capital for investments in Mozambique in general?
     - How would you assess the availability of domestic capital, that is capital from people or institutions in Mozambique?
       - What are reasons for this situation?
     - How would you assess the availability of foreign capital?
       - What are reasons for this situation?
   - To what extent are potential investors willing to spend their money on smart grids in Mozambique?

– What are reasons for this situation?

   – In which way does the potential of smart grids, to improve the quality of energy supply, influence the willingness of investors to spend their money for smart grids? If necessary, please differentiate between different investors, like donors, public institutions or private enterprises.

   – What are potential risks to smart grid investments? How strongly do they influence smart grid investments?

   – In your assessment, are economic decisions of potential investors of smart-grids in Mozambique rather long-term or short-term oriented? How does this influence smart grid-investments?

4. Electricity market

   – Who are the major players in the Mozambican electricity market?

      – How powerful are they?

      – How does this structure of market power influence smart grid implementation in Mozambique?

   – Economic actors constantly face costs for gathering information about new technologies, developments in the market, to inforce contracts, negotiate and plan. Do you think that in Mozambique, these costs will be different for smart grids than for conventional ones?

– How do these so called transaction costs influence smart grid implementation in Mozambique?

– What do electricity tariffs in Mozambique look like?

– Does this tariff structure play a role for smart grid investments?

– How does this influence look like?

5. Institutions, policies and regulation

– In your opinion, how does the performance of political actors (government, parliament, parties) in Mozambique influence smart grid implementation?

– How does the regulatory framework influence smart grid implementation?

– Mozambique´s receives development aid. How do donors influence smart grid implementation in Mozambique?

– Mozambique is a member of international organizations, such as SADC and African Union. How does this international network influence technology transfer, infrastructure investments and smart grid implementation in particular?

6. Implementation and acceptance of infrastructure

– In your opinion, under which circumstances is a smart grid infrastructure applicable in Mozambique?

– What problems might occur?

– In many parts of the country, there is no grid at all. In these areas, under which circumstances will investors directly start with a smart grid instead of a conventional one?

        – How is the state of internet coverage in Mozambique and what impact does this have on smart grid-implementation?

        – What is the state of research and development regarding smart grid-solutions that go with the Mozambican needs?

        – How present are the potentials of digitalization to actors in the Mozambican electricity market?

- To what extent do you think that smart grids can be professionally operated in Mozambique?

        – Why do you reach this assessment?

- In your opinion, in which quality will maintenance of smart grids in Mozambique be realized in Mozambique?

        – Why do you reach this assessment?

- Under which circumstances do you think that a smart grid-infrastructure will be accepted by inhabitants and consumers?

        – Do you expect differences in the acceptance of smart and conventional grids?

        – Why do you reach this assessment?

7. Final questions

- Is there anything I should have asked? Something else you would like to tell me?

Notes:

The categories under examination in the interviews are based on the potential drivers and barriers, derived in chapter 5.1.3. The guide leaves room for the respondent to present his or her specific knowledge but consequently confines the interviews to the topic of interest and to the expert´s area of expertise. Thus, off-topic explanations or statements about areas which are not part of the respondent's experience are avoided.

The interview guide contains key questions and additional questions that can be asked to specify the given answers if necessary. A guide does not necessarily have to be strictly followed if spontaneous alterations or digressions deliver relevant insights. An alteration can make sense if there is a need for specification, add-on questions are required to react to new insights or if unexpected changes in the conversation occur.

Interview guide for part two of the study (options)

1. Introduction
   – Thank the respondent for his or her participation
   – Ask if respondent accepts to be recorded
   – Ask if the interview can be published with the respondent´s full name or if he or she wishes to remain anonymous

2. General information
   – In what context is your enterprise/institution/organization dealing with smart grids?
   – Can you describe your position in your enterprise/organization/institution?
   – In what context are you personally involved in these projects or programs involving smart grids?

3. Central vs. decentral
   – There are different options for electrification, such as extension of the central grid, decentral mini grids, off-grid solutions and maybe further possibilities. In your expectation, how will each of these different options contribute to electrification in Mozambique in the close future? Do you expect a combination of different options or a concentration on one option?

– Will there be regional differences regarding the viability of each of these options?

– What makes this option or this combination of options more likely to be implemented than the others?

     – Possible specifications: Financing, political support, development cooperation, aspects of infrastructure, acceptance, complexity, conditions in the electricity market, tariffs and fees

– Do you think that this option or this combination of options is the most desirable way for electrification? Please explain.

– If there will be a combination of electrification options, which share would you broadly expect for the different options?

4. Digitalization

     – In your opinion, how can the central grid in Mozambique benefit from the advancements in digital grid management and communication technology, (such as automated load management, forecasting, consumer participation, remote monitoring and mobile payment)?

     – In your opinion, how can mini grids in Mozambique benefit from the advancements in digital grid management and communication technology.

     – In your opinion, how can off-grid solutions in Mozambique benefit from the advancements in digital management and communication technology?

- In your opinion, how strongly does each of the options benefit from digitalization? (which option benefits most, secondly…)
- Which potentials do you see for consumer participation in Mozambique?

5. Specific Regulation and political framework

- Does a mini grid-specific regulation exist in Mozambique? What does it look like?
- Possible specifications:

    - In Mozambique, are there rural electrification agencies that regulate decentral electricity generation and distribution?
    - Are there connection charges? If so, what do you think about their impact?
    - Is the price cap for power tariffs also compulsory for small power distributors?
    - What is the effect of the rigid tariff regime on mini grids?

- Does regulation account for the specificities of connected mini grids?

    - Which rules for the interconnection of national grid with mini grids exists?
    - Is interconnection relatively easy or not?
    - Possible specifications:

        - Is there a guaranteed interconnection when the national grid arrives, if certain basic requirements of quality are met?

- Are there clear financing responsibilities for the case of interconnection?
- Are there standardized procedures for interconnection?
- Is there a guaranteed power purchase from connected mini grids?
- Are there standardized power purchase agreements?
- Is there a guaranteed sale of back-up power to the mini grid when instability issues occur?
- Are there specific regulations for off-grid solutions in Mozambique? If so, what do they look like?

6. Final questions
    - Is there anything I should have asked? Something else you would like to tell me?

Notes:

The guide leaves room for the respondent to present his or her specific knowledge but consequently confines the interviews to the topic of interest and to the expert's area of expertise. Thus, off-topic explanations or statements about areas which are not part of the respondent's experience are avoided.

The interview guide contains key questions and additional questions that can be asked to specify the given answers if necessary. A guide does not necessarily have to be strictly followed if spontaneous alterations or digressions deliver relevant insights. An alteration can make sense if there is a need for specification, add-on questions are required to react to new insights or if unexpected changes in the conversation occur.